PELICAN BOOKS

A 75

WATCHING BIRDS

JAMES FISHER

JAMES FISHER

WATCHING BIRDS

PENGUIN BOOKS

HARMONDSWORTH · MIDDLESEX

FIRST PUBLISHED FEBRUARY 1941
REPRINTED 1941, 1946
REVISED EDITION
1951

TO

MY FATHER

❧

MADE AND PRINTED IN GREAT BRITAIN
FOR PENGUIN BOOKS LTD
HAZELL, WATSON AND VINEY LTD
AYLESBURY AND LONDON

CONTENTS

ᛉ

*The characters in this book are not intended to be
fictitious; they are all alive now.*

ILLUSTRATIONS

PREFACE TO THE FIRST EDITION

SOME people might consider an apology necessary for the appearance of a book about birds at a time when Britain is fighting for its own and many other lives. I make no such apology. Birds are part of the heritage we are fighting for. After this war ordinary people are going to have a better time than they have had; they are going to get about more; they will have time to rest from their tremendous tasks; many will get the opportunity, hitherto sought in vain, of watching wild creatures and making discoveries about them. It is for these men and women, and not for the privileged few to whom ornithology has been an indulgence, that I have written this little book.

Within the obvious limits, I have tried to make this account of birds up-to-date and scientific, and to illustrate some of what we know by pictures as well as text. Most of these pictures have been specially drawn by Geoffrey Salter, to whom I am very grateful. Two (Figs. 27 and 29) are by Hugh Cott, and are from his fine book *Adaptive Coloration in Animals*; I am very proud to have them. Mary Nicholson drew Fig. 24, and some (Figs. 9, 11, 16, 22, 23, 26, 31) I drew myself.

Some of the text is adapted from articles contributed to *Animal and Zoo Magazine*; I am glad to acknowledge this. I must thank, too, the editors of *The Handbook of British Birds*, David Lack, H. N. Southern, A. W. Boyd, P. A. D. Hollom, Dr Julian Huxley, Dr A. Landsborough Thomson, and the Controller of H.M. Stationery Office for permission to reproduce, adapt, or otherwise use material of an original nature. I hope, moreover, that David Lack, who read the book in proof, will not think his task entirely thankless.

I shall be very pleased if anybody who reads it becomes interested in birds.

JAMES FISHER

Oundle, November 1940

PREFACE TO THE THIRD PRINTING

RATHER over four years after this book first came out, 446 people had become new subscribers to the British Trust for Ornithology, as the result of reading it. I believe this is because the *study* of birds (a business somewhat different from an *interest* in them or an emotional *love* for them) concerns a very large number of the ordinary people who meet each other in the street and buy Penguins to read in trains. The B.T.O. organizes and undertakes field work, and its main object is study, and the making of new discoveries.

Many of those who joined the B.T.O. through this book have already taken part in inquiries and have contributed to making new discoveries. I know that some have got a great deal of pleasure out of so doing, and I need hardly add that I hope still more will join.

J.F.

Ashton, August 1945

PREFACE TO THIS EDITION

AFTER ten years at least 601 people have joined the Trust as a result of reading *Watching Birds*. Much has happened in Ornithology, and I have had to make quite a number of changes in the text to bring it up to date. But I have tried to keep as much as possible to the original plan.

In the first paragraph of my first preface (1940), I was rash enough to make some prognostications. As far as I can make out, these have come quite true.

J.F.

Ashton, December 1950

CHAPTER I

Introducing the Bird-watcher
to the Bird

❧

*My remarks are, I trust, true in the whole, though I do not pretend to say
that they are perfectly void of mistake, or that a more nice observer might
not make many additions, since subjects of this kind are inexhaustible.*

GILBERT WHITE, 9 December 1773

❧

The Bird-watcher. – Though they are by no means the most
ubiquitous, numerous, curious, or diversified forms of life, it
is nevertheless a fact that birds have had more attention paid
to them than any other corresponding group of animals.
Many attempts have been made to discover why so many
people like birds. Nobody has so far given a really satis-
factory answer, though when it comes, it will probably be
found partly from a sort of mass observation and partly from
the long and detailed history which bird-watching has
enjoyed, particularly in Britain.

All sorts of different people seem to watch birds. Among
those I know of are a Prime Minister, a President, three
Secretaries of State, a charwoman, two policemen, two
Kings, two Royal Dukes, one Prince, one Princess, a Com-
munist, seven Labour, one Liberal, and six Conservative
Members of Parliament, several farm-labourers earning
ninety shillings a week, a rich man who earns two or three
times that amount in every hour of the day, at least forty-six
schoolmasters, an engine-driver, a postman, and an up-
holsterer.

The reasons which make these men and women interested

in birds are, I am sure, very divergent. Some, when asked, can give no particular reason for their liking for birds. Others are quite definite – they like their shape, their colours, their songs, the places where they live; their view is the aesthetic one. Many of these paint birds or write prose or poetry about them. Still others, like several of the schoolmasters, are grimly scientific about them, and will talk for hours on the territory theory, the classification of the swallows, or changes in the bird population of British woodland during historic times.

The observation of birds may be a superstition, a tradition, an art, a science, a pleasure, a hobby, or a bore; this depends entirely on the nature of the observer. Those who read this book may give bird watching any one of these definitions; which is a sound reason why I should get down as soon as possible to the business of introducing the bird-watchers to the birds.

I should state my own position, however, to make it quite fair to the reader. My attitude towards bird watching is primarily scientific, or so I like to think. Those of you who want passages, purple or otherwise, on the aesthetics of bird watching, will not find them here.

The Bird. – Birds are animals which branched off from reptiles at roughly the same time in evolutionary history as did the mammals. Though both birds and mammals have warm blood, from anatomical considerations it is clear that these two branches are quite separate, and that the birds of to-day are more closely related to reptiles than they are to mammals. The parts of the world that birds live in are rather limited. Though birds have been seen in practically every area of the world from Pole to Pole and at heights more than that of Mount Everest, yet they have never been found to penetrate more than about 11 feet into the earth or about 100 feet below the surface of the sea; and every year some part of their lives has to be spent in contact with land, for birds cannot build nests at sea.

In the hundred and thirty million years or so in which

birds have existed as a separate class of animals, they have retained their basic structure to a remarkable extent. To avoid going into anatomical details, it must be enough to say that the same bones (that is, homologous ones) exist in the wings of humming-birds and ostriches, and in the feet of hawks and ducks. What has happened in the course of evolution is that these have been modified for different purposes. The wing of the humming-bird has become perhaps the most efficient organ of flight in the animal kingdom (the humming-bird is, as far as I know, the only bird which can actually fly backwards[1]); while that of the ostrich, which can no longer fly, is used as an organ of balance and of display. The talons of the hawk (Fig. 1), provided as they are with terrible claws, grasp its prey, and are often the weapons which cause the victim's death (rather than the beak); while the feet of the duck are webbed so that it can walk upon mud and swim.

Many are the ways in which birds are adapted to their surroundings, and to the lives which nature (or, if you like, evolutionary history) has taught them to live. Many interesting adaptations can be seen in British birds: so that they may climb trees and cling to them while hunting insects, the woodpeckers have feet with two toes pointing forwards and two aft, and the under-side of their tail feathers is rough where these rest against the trunk of the tree as the birds climb; their beaks are like chisels, for often they carve much of their nest out of the living wood, and their food habits take the great beaks under bark and into crevices in their search for insects. In North America one group of woodpeckers has gone even further, not so much in adaptation as in the habits produced by such an adaptation. They have deserted insect catching for sucking the sap of the trees – hence their name of sap-suckers.

Many birds are beautifully adapted to their surroundings at the earliest stage of their lives: the egg. The eggs of the

1. After this was written a note by A. B. Williams appeared in *The Auk*, vol. 27, p. 295 (1940), showing that the Acadian flycatcher can 'perform this feat very neatly and apparently very easily.'

A. The raven is a typical perching bird; its feet are adapted to clinging to trees and
rocks and to the carrion on which it feeds.

B. The gyr-falcon is a powerful bird of prey, catching animals up to the size of ptarmigan
in its talons. These are tremendously strong and muscular, and armed with sharp and
prehensile claws.

C. The cuckoo spends a good deal of its time perching and is a good climber, too. Two
of its toes point forwards and two backwards (a condition known scientifically as
zygodactyle); this is an efficient adaptation to perching.

FIG. 1. The feet of certain birds, showing how they are adapted to
different modes of life.

ringed plover and of other shore birds often resemble very
closely the pebbles among which they are laid (see Fig. 27,
p. 122); those of birds which build open nests are often
coloured with a protective pattern of pigment. But it is a
biological economy to lay an egg without pigment, where
pigment is unnecessary. The eggs of woodpeckers and owls
are white when they are laid, and coloured later on only by

D. The green woodpecker spends most of its life climbing up the boles of trees (though it has lately shown a tendency to take to the open countryside). To this end its feet are arranged in the zygodactyle condition, and as it climbs it gets further support from the under-side of its tail, the feathers of which are rough.

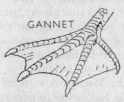

E. The gannet hunts its prey (fish) by diving upon it from a height; once under the water it swims after it, using both feet and wings. It swims well on the surface of the water, and like almost all birds of the ocean, has webs embracing all four toes, the hind toe pointing forward for the purpose. The middle toe has a pectination for combing the plumage.

F. The great crested grebe is a water bird that has palmate rather than webbed feet; these are probably just as efficient. Palmate feet have been evolved separately by more than one unrelated group of birds, as have webbed feet, for that matter.

FIG. 1, *continued.*

dirt, and these birds nest in holes where their eggs cannot be seen by enemies. Oddly enough, the egg of the puffin still has grey or red spots, like faint copies of those of its cousin, the guillemot, which rather suggests that the puffin has only recently (in the evolutionary sense) taken to nesting in burrows.

When we speak of adaptation of the sort which we have

just described, we must remember that our attention is likely to be drawn to the more remarkable examples of adaptation to special needs and away from the fact that practically every organ and part of an animal is adapted (directly or indirectly, efficiently or inefficiently, in greater or lesser degree) to the animal's surroundings, or to its demands at some period of its total or daily life. A bird is what it is partly because of the immediate environment – since food or weather may affect its colour or size; partly because of heredity – which provides machinery, producing in its turn variations on which natural selection can act; partly because of chance – since chance appearance of such new variations as are ignored by natural selection may determine some of the characters of an individual or even of an isolated group of individuals. We must remember that not only special kinds of beaks but beaks in general are adaptations. A wing is an adaptation for flying. An egg is an adaptation for nourishing the young. Adaptations, in birds as in other animals, must be regarded in terms of the perspective of the ages, the dynamic of evolution, and the complicated mechanism of heredity. We may regard any characteristic as adaptive if it can be explained as fitting the bird to its mode of life.

How a Bird is Built.—The skeleton of a bird (Fig. 2) is, in general plan, the same as that of other vertebrate animals. Essentially, it consists of a strong box made up of many bony elements, the head, first, which is attached to a column of strong articulating vertebrae, the backbone. To the axis provided by the head and vertebrae are attached various appendages; to the head the jaws; to the lower vertebrae of the neck from one to four 'floating' ribs; to the vertebrae of the back, ribs which curve forwards to join the huge breast-bone, forming a strong case inside which the lungs and heart lie, and to the outside of which is attached, towards the back, the shoulder girdle; to the remaining vertebrae the hip girdle. The shoulder girdle supports the elements of the wings, the hip girdle those of the legs.

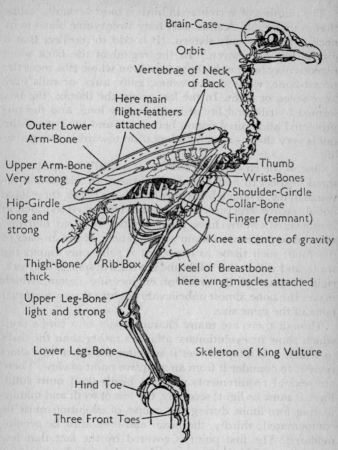

Brain-Case

Orbit

Vertebrae of Neck of Back

Here main flight-feathers attached

Outer Lower Arm-Bone

Upper Arm-Bone Very strong

Hip-Girdle long and strong

Thigh-Bone thick

Rib-Box

Upper Leg-Bone light and strong

Lower Leg-Bone

Hind Toe

Three Front Toes

Thumb

Wrist-Bones

Shoulder-Girdle

Collar-Bone

Finger (remnant)

Knee at centre of gravity

Keel of Breastbone here wing-muscles attached

Skeleton of King Vulture

FIG. 2. Skeleton of king vulture (an American bird of prey which, as a matter of fact, is not a true vulture), with wings folded in the normal position of rest. The main bones are indicated, though their technical names are mostly avoided.

The number of vertebrae in birds is most variable, unlike that in mammals. A swan may have twenty-five bones in its neck, a sparrow only sixteen. (It is odd to recollect that a giraffe has only seven.) In the region of the back some pigeons may have only three vertebrae whose ribs meet the breast-bone, while ducks, swans, gulls, auks, or rails may have seven or eight. In the last group the thorax (the box formed by ribs and breast-bone) is very long, and the un-protected abdominal region between it and the base of the tail is very short – a useful adaptation towards a swimming life.

One of the chief characteristics of a bird's bones that we should expect to find is that of lightness, an adaptation to-wards flying. Though birds may have large bones, these have not a massive structure. They are not reinforced so much as cantilevered. Many have great cavities with no marrow, some of which may contain prolongations of the air-sacs of the lung; such tissue as is present is solid only round the walls, and within this solid tube is a mass of spongy network, rigid because of its excellent engineering design, which makes the bone almost unbelievably lighter than a mammal bone of the same size.

Though there are many characteristics of a bird's skull which show its evolutionary affinities rather than the tasks to which it is adapted, yet it is better for us, in this short review, to consider it from an adaptive point of view. There are several requirements which a bird's skull must fulfil. First, it must be light; secondly, the loss of teeth and manip-ulating fore-limbs during the course of evolution must be compensated; thirdly, the brain and eyes must be accom-modated. The first point is covered by the fact that few skull-bones of birds are more than plates and struts; the second by the high development of the horny beak, which has to act as a limb, and is, in practice, a very effective one; the third by the huge orbits, which leave only a very thin partition between the eyes on each side. These restrict the rest of the brain to the back of the skull – which is accordingly broad. That part of the brain which deals with smell (of which birds have very little sense) is very much reduced; that which deals with co-ordination of movement and balance is large; that which deals with what are sometimes called the 'higher functions' is small.

The greatest differences between the skeletons of birds and other vertebrates are found in the skull and in the pectoral girdle. The latter corresponds to the shoulder-blade, collar-bone, arm, forearm, and hand of man. At only one place is this girdle attached to the bones of the main trunk; it is

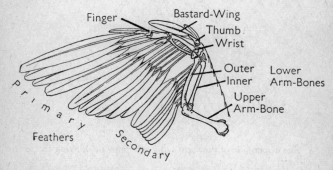

Fig. 3. The under-side of the right wing of a pigeon (after Marshall), showing the arm-bones and the origin of the important flight feathers.

where the front end of the shoulder-blade is attached, on each side, firmly to the fore-part of the breast-bone. The two shoulder-blades are braced together across the front by the collar-bones, which are fused in the middle to form what we call the wish-bone. Behind, the free ends of the shoulder-blades are bound by stiff and strong ligaments to the ribs and to the vertebrae of the back.

It is clear therefore that when the wing is working the strain comes directly on to the box formed by vertebrae, ribs, and breast-bone which we have called the thorax. This box has therefore to be extremely strong. It is. At the back the vertebrae are fused together, at the sides the ribs are bound to each other by further ligaments, and in front the breast-bone is built on engineering principles that combine strength with great lightness and a keel for the attachment of the powerful muscles of the wing.

This is the basis upon which we can build our wing. The part that flaps (Fig. 3) is composed of three units. Attached to the shoulder-blade by a ball-and-socket joint is the upper arm-bone, a single strong rod. To the other end of this are joined the two bones of the forearm. To them is joined the

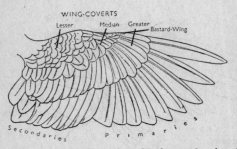

FIG. 4. The upper-side of the right wing of a pigeon, showing the groups of feathers.

third unit, several small bones, largely fused together, repre-senting the mammalian hand.

The muscular force which works the wing is derived al-most entirely from a huge plate of muscle that is attached from the wish-bone in front to the base of the breast-bone below and the shoulder-blade behind. The fibres of this great muscle converge to an insertion on the upper arm-bone (the humerus). It is the humerus, then, which takes the strain of the wing beat, though the feathers of the wing (Fig. 4) are carried not on this but on the other bones of the arm. The secondary feathers are attached to the ulna, the lower and thicker of the two bones of the forearm; these feathers play the major part in propulsion. The primary feathers are attached to an elongated wrist, and to two bones which correspond to the index finger; these play an effective part in steering. Thumb and third finger are present in rudimentary form, the former bearing a bunch of feathers often called the bastard wing, which appear to have the aeronautical function of a wing slot. The other fingers are absent.

All other attempts at flight by vertebrates (and there are upwards of half a dozen) have involved the development of flaps of skin, stretched, as in the pterodactyl, from the little finger to the feet, or, as in the bat, over all the outstretched fingers to the feet (Fig. 5). In no case except birds have feet been kept out of the mechanism; besides which, skin is clumsy to fold compared with feathers. So it can be seen that birds have a considerable advantage over other flying vertebrates. The extent of this advantage can be gauged from the fact that pterodactyls are extinct and bats restricted in range and habits, whereas birds are almost everywhere.

As birds have been able to keep their feet out of their flying mechanism, these organs have been left free for the play of evolutionary adaptation. But because birds can fly, there are some considerable problems to be got over before they can run. The great pectoral muscle that works the wing may weigh, in the pigeon, over a quarter of the weight of the whole body. The centre of gravity of a flying bird lies always considerably in front of the joint between leg and body. Hence, if the bird is to stand upright, the girdle of the hip must clasp the vertebrae in a vice-like grip. This, of course, is what it does. In front of and behind the socket into which the head of the thigh-bone inserts, the central part of the hip-girdle is elongated and fused closely with the vertebrae. It will be remembered that the vertebrae of the back were fused together the length of the thorax, to give a strong basis to the working of the wing. The last one or two of these rib-bearing vertebrae take part also in the hip girdle, and are clasped by the broad iliac bone. Below them, perhaps ten or eleven more vertebrae are fused with the girdle, and beyond these two or three more little ones may project; these are the last of

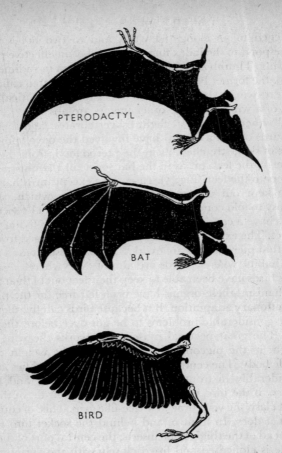

FIG. 5. Comparison between the wings of pterodactyl, bat, and bird. Neither pterodactyl nor bat has been able to keep its legs out of the flight mechanism; the wing of the former is stretched from little finger to leg and tail, that of the latter from four fingers (not the thumb) to leg and tail. The bird's wing does not involve its legs or tail, and the flight feathers are attached to one of the lower arm-bones and to the reduced bones of wrist, hand, and fingers. (Figure after Huxley, Wells and Wells.)

the caudal or tail vertebrae, and represent what there is of a true tail in modern birds. So the total result is that the bird has a very rigid back from the base of its neck to its tail, a back whose rigidity is secured by the fusion of vertebrae to each other or to the hip girdle, and which itself secures for the bird strong flight and the power to support itself firmly upon its legs.

What appears to be the thigh of a bird is in reality its shin, the leg below the true knee. The true thigh-bone is usually short, and runs almost horizontally forwards to the knee, which may lie closely applied to the surface of the body just about under the centre of gravity.

Running downwards and slightly backwards the true shin (apparent thigh) is a long bone; all that is left of the second leg-bone found in most land vertebrates is a sliver applied to the surface of the main bone. The next member runs downwards and forwards, and is a single bone derived from elements originally part of the ankle and the upper ends of three of the toes. The toes themselves are never more than four in number; the first (big toe) is nearly always turned backwards when it is present. The fifth (little toe) is entirely absent.

It can be seen that the bird's leg, composed as it is of three single rigid bones, jointed together, provided with toes at the end, is ideal for its purpose, and can be adapted for running, standing, taking landing strain, and catching prey. The knee- and ankle-joints of the bird both lie some distance from the hip-joint and toe-joints, and work in opposite directions. Thus they constitute perhaps the most effective shock-absorbing mechanism in the animal world.

The Soft Parts.—Anybody who has prepared a fowl for the table will know, and those who have carved one will suspect, that the 'guts' of a bird (Fig. 6) are contained in a cavity stretching from rib box to hip girdle, and covered very largely by the downward projection of the breast-bone. This cavity is generally known as the abdominal cavity, and in it the long continuous tube of the viscera is slung. Food is taken in by the beak, laced with saliva from glands in the mouth, and squeezed down the throat by muscles at the back of the mouth. In the throat region the food tube is known as the gullet or oesophagus. It may be expansible and used to store food, or it may have a large special bag attached to it for storage; if present, this is known as the crop. Crops may sometimes be used as 'udders'; those of pigeons secrete a

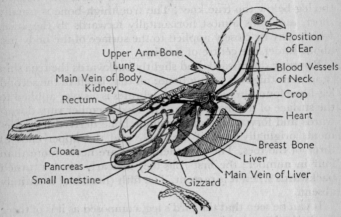

Upper Arm-Bone
Lung
Main Vein of Body
Kidney
Rectum
Cloaca
Pancreas
Small Intestine
Gizzard
Position of Ear
Blood Vessels of Neck
Crop
Heart
Breast Bone
Liver
Main Vein of Liver

FIG. 6. Organs of the pigeon; the main soft parts are indicated. Note the very large size of the heart, and the crowding of the intestines into the lower part of the body. (After Marshall.)

milk-like fluid used to feed the young. Sometimes they may be used as extra gizzards; that of the hoatzin has muscular walls which can squeeze the juices out of its food.

The gullet continues through the thorax, passing behind the heart, to reach the abdominal cavity, where it immediately broadens out into a large bag, the stomach. Usually the first part of this has walls supplied with glands which secrete acid, in order to digest the food, and the after-part is lined with muscle and developed as the gizzard. Many birds which live on grain or on hard vegetables have large gizzards and frequently swallow small stones, which are retained in the gizzard as an aid to crushing. Others which live on flesh have no gizzards at all, but have glands over the whole of the stomach.

After the stomach the food tube is continued as a loop known as the duodenum. Into this run three ducts from an organ which is slung in this loop, and from the wall of the stomach; this organ is the pancreas, and down the ducts run secretions which change starch to sugar and break proteins and fats down to simpler forms. Ducts also run into the duodenum from the liver, a huge organ which may be even larger than the stomach, and which often surrounds it, with its top surface closely applied to the diaphragm, the broad muscular wall which divides the cavity of the thorax from that of the abdomen. The secretions from the liver into the food canal, which often pass through a reservoir known as the gall-bladder, are mainly waste products, but contain a few elements useful in digestion. The main function of the liver seems to be the storing of sugar, and it acts as a kind of governor in keeping the constitution of the blood constant.

After the duodenum we reach the small intestine proper. This is often very long, especially in vegetable-feeding birds. In it digestion continues, and the main part of the absorption of the digested products takes place. It loops and coils about the abdominal cavity, each loop being held in place by a membrane attached to the back wall, and practically fills the cavity. Finally, it joins the rectum, a wide short tube which leads to the cloaca. At the junction of the small intestine with the rectum we find one or two blind tubes branching sideways. Sometimes these are quite large, and in them it is possible that matter like cellulose is broken down by bacterial action and rendered fit for absorption.

The final passage, the short cloaca, communicates with the exterior, but before this it receives the excretory ducts from the kidneys and the reproductive ducts from the testes or ovaries. Thus the cloaca is a general-purpose opening; it is both excretory and reproductive, and is the only such open-ing in the lower part of the bird's body.

The Heart and Blood. – Though the heart of birds is built in a different way from that of mammals (it has more affinities with the reptile heart), it is four-chambered, and apparently is just as efficient, since in it oxygenated blood is effectively separated from the de-oxygenated blood which has delivered up its oxygen to the body. Birds have a more rapid heart-beat than mammals, and their temperature is considerably higher – from 100° to 114° Fahrenheit. Like mammals, they keep the same body temperature all the time; that is, the temperature of their body is constant and independent of the temperature of their surroundings. As in mammals (e.g. bats), so there are, in birds, some exceptions to this rule. Under certain conditions of the environment, such as the cold nights of the high Andes, the maintenance of body temperature by small birds like humming-birds (which have a relatively large sur-face in proportion to their weight) becomes difficult or im-possible. Some humming-birds, under these circumstances, become torpid. At first their breathing becomes violent, and they lose the power of flight. When touched, they may make a peculiar whistling sound. When completely torpid they are quite rigid, the head is pointed upwards, the eyes are closed, the breathing appears to stop. It may take the birds about half an hour to recover from this state. At least one kind of bird, the poor-will of western U.S., can truly hibernate, and when it does so its temperature falls considerably.

Thus under certain exceptional circumstances birds can revert to the reptile-like condition of having a body tem-perature that is not constant. In adult birds this condition, then, is very rare, is only found in certain birds, and, when present, is an adaptation to special circumstances of the environment. On the other hand, the young (in the nest) of most birds show a variable body temperature. This is because the mechanism of temperature control does not get into full working order until most of the feathers have reached their adult development. Hence the care which many birds take, by incubation and by shading from the sun, in protecting their young from excessive cold or heat.

The rapid heart-beat of the bird is a necessary adaptation, not only towards keeping up its temperature, but towards supplying the huge demands of the flying muscles. These require sugars, and oxygen to 'burn' them with, so as to produce the large amount of energy needed. And the blood has to carry the necessary supplies from the liver – the sugar storehouse – and the lungs. The heart itself is large, and takes up more space in the thorax than does the heart of a mammal; its walls are very thick and muscular, for it has a lot of work to do in pumping the blood round the body – blood which carries, besides sugar and oxygen, the waste gas carbon dioxide (the product of the burning of the sugars), other waste products, proteins for body building, fats for storage and more energy, the secretions of the ductless glands (chemical compounds often called hormones which act as 'messengers' and whose task is largely parallel with that of the nervous system), and various bodies concerned with the destruction and neutralization of bacteria and poisons.

The Lungs. – A bird's windpipe opens at the back of the tongue and runs down the throat to the top of the thorax. Here it broadens out into the song-box. This consists of a main chamber, with a bony band inside. This band has two processes attached to it on which a membrane is stretched; there may be other membranes present as well, and together these act as vocal cords, producing sound and song. From the bottom of the song-box two tubes branch off, one to each lung. These tubes feed not only the lungs but five air sacs which are often large. These sacs are not adaptations towards buoyancy so much as mechanisms for increasing the amount of air available inside the bird for use – very useful if the bird is a great songster or a great diver. The sacs also fill the lungs with air directly after an out-breathing, so that air is available for respiration in the periods before it can be breathed in from outside. And since birds cannot sweat, a lot of the water exchange which helps to keep their tempera-

ture constant takes place over the inner surface of these
sacs.

The Reproductive Organs (Fig. 7). – In the abdominal cavity,
closely applied to the back wall on each side of the mid-line,
lies a pair of flattish, lobed organs. These are the kidneys,
and from the middle of each a slender tube runs direct to the
cloaca. The kidneys collect waste products from the blood
and excrete them in liquid solution through the ducts.

Most female animals have a pair of ovaries slung from the
wall close to each kidney. Most birds have (for some reason)

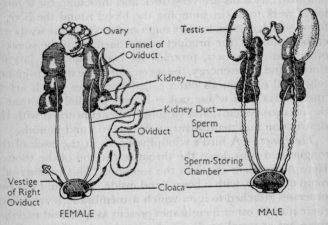

FIG. 7. The reproductive organs of a male and female bird. (After
Thomson.)

only the left; if the right is present, it is nearly always a ves-
tige. The ovary looks like a bunch of grapes. A long, thick,
coiled tube which runs to the cloaca on the left side has at its
upper end a great funnel which opens directly into the
abdominal cavity near the ovary. This is the oviduct.

When the breeding season arrives, a chain of events is set
in motion inside the bird. Certain ductless glands begin to
pour their secretions, often under the influence of a special

ductless gland in the region of the brain – the pituitary body – which appears to act in some ways as a master gland. Changes begin to take place in the ovary (which is a ductless gland in its own right, producing hormones like the others), and some of the grapes on the surface of the bunch, which are in reality unfertilized and immature eggs, begin to take in yolk and get bigger. The eggs are ripening.

In a male bird organs can be seen attached to the back wall of the abdominal cavity in much the same place as the ovary is in the female. In the male, however, they are paired and of equal size and function, lying, as the left ovary does, near the top of the kidneys. These are the male reproductive organs, and they are oval bodies, tiny in the off-season but huge at breeding-time; they consist of a part known as the testis, in which the sperms and some hormones are produced, and another part mainly concerned with storage of the male fluid. From each organ runs a narrow duct, of about the same size as the tube from the kidney, to the cloaca.

When birds copulate, the female generally takes up a crouching position and lifts her tail. The male mounts on her back and applies his cloaca to hers. By a muscular action which is probably out of the male's conscious control, a dose of male fluid is then pumped into the female's cloaca from little widenings at the bottom of the sperm-ducts, where it has been stored. Some male birds which might otherwise have difficulty in securing efficient passing of the sperm, such as ducks, which often copulate on the water, have a special muscular penis attached to their cloaca which fits into that of the female. Not many birds, however, are provided with such an organ.

Development. – The sperms of the male birds, of which there are many thousands in a dose of fluid, are living things with an independent existence, and have many of the properties of separate organisms. One of these properties, that of swimming against a current, now comes into play; and the sperms make their way, against the slow movement of the secretions from the walls of the oviduct, up this tube.

They may meet an egg or eggs either at the top of the oviduct or in the oviduct itself; when they do, they fuse with the egg. As soon as a sperm has entered an egg the egg's surface changes chemically and no other sperms can get in. The egg, once fertilized, rapidly begins to divide. It already has a supply of yolk and consists of a dividing, developing group of cells above and a mass of nutritious yolk below.

The next step is for the yolk with the developing embryo upon it to be wrapped in a safe and useful parcel. This is done by cells in the walls of the top of the oviduct, and the first wrapping consists of an albuminous substance which, in the final full-sized egg, can be seen as a sort of curly part of the white between the yolk and the ends of the egg. When the egg is laid and the shell is hard, this curly part acts as a kind of spring shock-absorber. As the egg passes farther down the oviduct, more of the white of egg is secreted and deposited round it. Still farther down the tube are cells which secrete the chalky shell, and farther on still are more cells whose job it is to produce pigments to make the typical pattern on the egg. The egg is finally laid, broad end foremost, by contractions of muscles in the wall of the cloaca.

The embryos of animals more simple than the hen, like those of frogs, for instance, develop by division of the fertilized egg and rapidly form a little round mass which then becomes something like a yolk-filled cup. The embryos of birds (Fig. 8) cannot develop in this way because of the great size of the mass of yolk. So instead of growing as a roughly ball-shaped mass, they start life as a plate of cells applied to the top of the yolk. To begin with, this plate is streak-shaped and very soon after the egg has been laid the first cells of the nervous system can be seen in the mid-line. By the end of forty hours the new animal is over a quarter of an inch long, is more sausage-shaped than streak-shaped, and is beginning to separate a little from the yolk. This separation is most pronounced in the front third of the growing body. This part is going to be the head, and already swellings can be seen where the eyes are to be; down the rest of the body little round dots

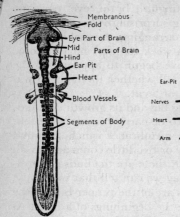

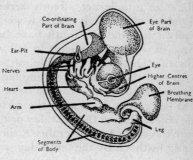

A. Thirty-six hours after the egg is laid; back view.

B. Five days after the egg is laid; side view. By now the limbs have begun to appear, the breathing-membrane has developed, and the eyes are well advanced.

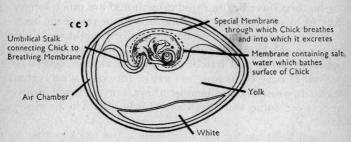

c. Nine days after the egg is laid; view of the embryo in the egg and its surrounding membranes. By now the chick is well advanced and has really begun to look like a young bird. (After Marshall and Hurst.)

FIG. 8. Stages in the development of the chick.

indicate the segments of the growing bird. By the end of four days from laying we can distinguish brain, eye, ear, and heart, the last as a sort of bag surprisingly near the head. Plates of muscle are already arranging themselves in the segments of the body. After a week the eyes have become really large and the limbs have begun to appear. All the same, it would be difficult to tell whether the embryo was that of a chicken, a crocodile, or a cat.

By now the chicken is getting big and its growth demands a supply of oxygen. This has to be carried in blood-vessels, and large ones develop in the membranes which surround the chick. During the next day or two little cones appear all over the surface of the embryo; these are the buds of developing feathers. By the tenth day we can tell that we have a young bird, because the limbs are of the proper bird shape and structure, and we can see the beginnings of a beak. At the end of a fortnight quite a lot of real bone has been laid down in the skeleton in place of soft cartilage, and the muscles are well developed, the beak has been provided with an egg tooth with which the chicken can break its way out when the moment of hatching comes about a week later, the feathers have begun rapid growth and are much longer, and the head is less huge in proportion to the rest of the body.

The developing bird we have described might well have been the chicken of the domestic hen. The period of time between laying and hatching is known as the incubation period. It may be as long as forty or fifty days with certain sea birds, ostriches, and brush-turkeys. For the hen it is three weeks; for some small perching birds only ten or eleven days.

When the time for hatching comes, the chick pushes its beak into the air chamber at the broad end of the egg and takes its first fill of air into its lungs. Up to now its oxygen supply had come from the air diffusing through the walls of the shell and received by the network of blood-vessels outside the chick itself. Having filled its lungs with air, the chick then strikes the egg shell and breaks it. After an hour or two it may have broken the broad end of the egg off completely and may have struggled into the outside world.

In general there are two kinds of chickens – those of nidifugous and those of nidicolous birds. Nidifugous birds can run as soon as they are hatched and are hatched at a comparatively advanced stage. Game birds, wading birds, and ducks have nidifugous chicks, all very active and capable of running, swimming, and hiding. The most striking example of a nidifugous bird is the young of the brush-turkey. The egg of this bird is incubated, not by any parent, but by the heat of fermentation of a pile of decaying vegetable matter scratched up by the male bird. The young often have to find their own way out of the pile of stuff, and as soon as they have managed to do this, they can flutter off the ground on wings at a remarkably high stage of development. On the other hand, birds which nest in holes and which build complicated and comfortable structures can afford to have their young hatched at an earlier period of development; these are, then, nidicolous. A newly hatched tit or blackbird has scarcely any developed feathers at all and is naked, ugly, and helpless.

CHAPTER II

Arranging the Birds

❦

Faunists, as you observe, are too apt to acquiesce in bare descriptions, and a few synonyms: the reason is plain; because all that may be done at home in a man's study, but the investigation of the life and conversation of animals is a concern of much more trouble and difficulty, and is not to be obtained but by the active and inquisitive, and by those that reside much in the country.

GILBERT WHITE, 1 August 1771

❦

Species. – There are about eight thousand five hundred species of birds in the world, according to the highest authorities. This is really quite a small number; there are over a hundred thousand kinds of molluscs and about three-quarters of a million kinds of insects (of which a third are beetles).[1] Yet there are probably more ornithologists than there are malacologists or entomologists in the world.

It is likely that bird-watchers are able to satisfy their keenness so well partly because of the interest taken in birds by taxonomic zoologists. Taxonomic zoologists usually inhabit museums and spend most of their lives arranging and sorting skins and skulls. These painstaking people have one of the most difficult tasks in the whole of zoology, for in their arrangement of their animals in a natural classification they have to bear in mind very many different principles. When a collection turns up from some part of the world that has not been very well worked, the taxonomist has to be an historian, an evolutionist, a geographer, an anatomist, a bibliographer,

1. Entomologists are still describing about ten thousand new species of insects every year.

and often a mathematician before he can be justified in assigning his specimens to their proper species or if necessary to a new species. Even then, if he has all these qualities, he may often have to rely on an intuitive flair for diagnosis.

In the Linnaean system every animal bears two names, a generic one and a specific one. The same generic name is not permitted by the international rules of nomenclature to occur twice in the animal kingdom, though specific names may often be repeated. Once a name has been given to an animal by an authority in a recognized scientific journal or some other such printed paper, it sticks unless somebody can later prove that a name has already been given to the animal concerned or that the animal does not merit being classed as a new species. This rule usually works very well and saves a great many mistakes and inconveniences. Sometimes, however, it has awkward results. For instance, the names of the British song-thrush have been changed several times in the last twenty years because of new discoveries about the dates of publication of the original papers in which it was described. The Manx shearwater is called *Puffinus puffinus* and the puffin *Fratercula arctica*. This seems curious to British bird-watchers, but has to be, since the name *Puffinus* was originally given to the shearwater.

For many years now zoologists have been concerned with the definition of the word 'species'. It should be realized that most modern taxonomists have two main points of view when deciding what a species is. They have to look at their animal, perhaps first from the point of view of practical convenience, and secondly they have to look at it from the evolutionary aspect. What they like to have as a species is a collection of animals which their colleagues would readily agree to be such, and which, when met with alive in the field, would have some reality to the animal watcher. Though at least fifteen different but complementary definitions of a species can be given, most of which are not enough by themselves but are very useful when taken together, it is often extraordinarily hard to decide whether a population of animals, including birds, really is all of one kind. In spite of these difficulties, I think most workers will agree that a species is a real thing and that the word describes actuality as well as degree.

If we are to define the word 'species', we have to do it in two ways, by stating the circumstances under which two creatures can be said to belong to the same species, and those under which they can be said to belong to different ones. Before we do this we should also state that (in birds particularly) a further subdivision is frequently used – the subspecies – which describes races which are separated geographically, but which do not themselves merit the rank of species. Let us now set out some definitions.

Two animals belong to the same species if such is the opinion of a competent systematist. This may appear to get you nowhere, but actually should get you inside a museum to ask somebody. It is not a cynical definition.

Two animals belong to the same species if they inhabit the same area, have the same tradition of structure, colour, voice, and habits, and tend to breed with each other rather than with similar animals which do not have that tradition.

Two animals do not necessarily belong to the same species if they interbreed in the wild. There are many examples of distinct species which have increased their range in the course of evolution so as to overlap. In the region of overlap they may interbreed, producing a mixed or hybrid population. Nevertheless, this does not mean that they are of the same species.

Two animals belong to different species if, when inhabiting the same area and having much the same outward form, they avoid interbreeding by devices of colour, voice, or display, or by selecting different habitats in that area. As we will see in a later chapter, chiffchaffs and willow-warblers are very similar and are quite hard to distinguish even in the hand. In nature they remain distinct species because of their distinctive songs (by which they recognize each other) and probably by slight differences in their choice of habitat.

A great range of size, colour, and geographical distribution does not destroy the reality of a species. Thus, if you
compared a puffin from Spitsbergen with one from Brittany
you might easily suggest that they were of different species
until you had seen a collection of birds intermediate in size
from the intervening coast of the North Atlantic. As a matter
of fact there is a beautiful gradation in size, increasing from
south to north. Taxonomists have given names to the populations in various regions up the coast, and these populations
can all be arranged in order of size and latitude. Not one of
these populations could be called a different species, but
they can be quite suitably called different subspecies or geographical races. It would be suitable to go even further and
say merely that there was a gradation or cline in size from
south to north and use, as Dr Huxley suggests, only the
names of the most southerly and northerly races (while,
of course, still pointing out about the cline between the
two).

Sometimes a population of animals may have been originally spread out over a large space. In the course of time for
various reasons this population may have been broken up
into isolated groups. As time goes on these groups, breeding
among themselves, may show tendencies to differ. As soon
as the average difference between one group and the next
becomes consistently and obviously greater than the average
difference between one individual within a group and
another, we are justified in calling each of these groups a
subspecies. When these differences become really great, we
are often forced to go further and call the groups different
species. Many examples under this heading are found on
islands. In Britain there are quite a lot of island forms of
birds. Thus there is a mainland wren, a Shetland wren, a
Hebridean wren, and a St Kilda wren (Fig. 9). There are
distinct Faeroe and Iceland wrens also, and the winter-wren
of America belongs to the same species, having a dozen
described subspecies in the northern part of that continent.
When the St Kilda wren was first described, it was given the

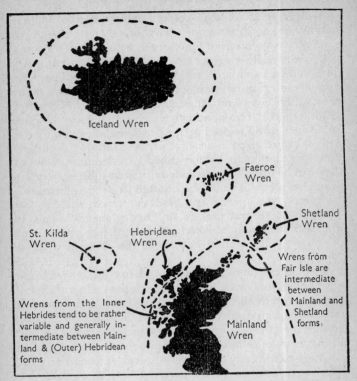

Iceland Wren

Faeroe Wren

Shetland Wren

St. Kilda Wren

Hebridean Wren

Wrens from Fair Isle are intermediate between Mainland and Shetland forms

Wrens from the Inner Hebrides tend to be rather variable and generally intermediate between Mainland & (Outer) Hebridean forms

Mainland Wren

FIG. 9. The subspecies of wrens in Great Britain, the Faeroes, and Iceland.

Mainland: *Troglodytes troglodytes troglodytes.*
Shetland: *Troglodytes troglodytes zetlandicus.*
Hebrides: *Troglodytes troglodytes hebridensis.*
St Kilda: *Troglodytes troglodytes hirtensis.*
Faeroes: *Troglodytes troglodytes borealis.*
Iceland: *Troglodytes troglodytes islandicus.*

In size from the largest to the smallest, they run Iceland – Faeroe and St Kilda – Shetland – Fair Isle – Hebrides – Mainland. It can be seen that this is a straight north-to-south gradation, St Kilda being the only exception; however, conditions on St Kilda are rather unusual – the population of wrens is small and isolated and the weather is severe – so we can excuse the St Kilda birds from rigidly following the rule.

rank of a full species, but zoologists have now decided that it should only be called a subspecies. Perhaps in a thousand years or so it may become even more different from the mainland form and may be worth calling a species again. The borderline between species and subspecies is (in the case of island forms) extremely difficult to draw.

I do not think it is worth while for the ordinary amateur field ornithologist or bird-watcher to trouble very much about the definition of the species he is working with. In Britain at all events the work has already been done very intensively, and though there is still a bit of a battle going on between those who like to recognize similarities between races of birds (lumpers) and those who like to recognize differences (splitters), their rivalry need not concern us very much. As far as British birds are concerned, the amateur can rest assured that the list of species is in the hands of a very capable committee of the British Ornithologists' Union. These learned gentlemen sit once or twice a year and discuss very carefully all the additions and subtractions necessary for the British list and all the changes of name that bibliographical and anatomical research have made necessary. If the reader wants to know the exact names of the species on the British list he should consult the British Museum List or the new *Handbook of British Birds*. The additions and corrections are published periodically in the *Ibis*, which is the journal of the British Ornithologists' Union, and if the bird-watcher uses these lists and with their aid examines his experience in the field, he will get a better idea of the reality of species and subspecies than any series of random theoretical definitions could give him. It is important nevertheless for him to realize that improvements in our knowledge of classification can come from his own work. Birds are notoriously hard to classify properly, and to help themselves in their task zoologists are having to depend on every new source of ideas that they can touch. To-day the systematist has to consider, not merely things like measurements, structure, and colour, but also voice, habits, geographical position, migration, population, type of nest, and egg.

Variation. – No two animals are alike. It is impossible to say that two animals are truly the same from the zoological point of view, except in the rare case of identical twins (and this is not really a safe case). We have already traced some general trends of mass variation, such as the trend in the puffin towards bigness as it gets north; we could give many more examples, thus, birds from moist regions tend to be darker than those from dry regions, and in cold parts of the world exposed parts tend to be covered with feathers. But this kind of organized and correlated variation is very different from individual variation. Sometimes certain individuals of a species are very different from the norm; for instance, they may lack pigment altogether, in which case they are termed albinos, or they may have excess of or altogether lack some but not all of the normal pigments. Often these variants are called sports, or mutants; study of them has often told us a lot about the laws of inheritance in birds. Sports or mutants are rare things, and those of extreme kinds probably play very little part in the evolution of a species.

Sometimes, however, variants appear very regularly, so that there is an apparent (and often a real) balance of numbers between variant and norm. Such a variant is present in our common guillemot, for certain individuals have a white 'bridle' or 'spectacle' circling and running backwards from the eye. Among the breeding stocks of guillemots in Britain, the bridled birds (Fig. 10) increase from under $\frac{1}{4}$ per cent of the total population in the south of England up to 26 per cent in Shetland. Beyond this northern point of Britain the percentage increases still further until it reaches 53 per cent in the south of Iceland and probably more on Bear Island. Clearly there is a definite gradation in space of the proportion between bridled and normal guillemots. This gradation, or 'cline' as Dr Julian Huxley might call it, is not perfectly continuous from north to south; thus there are a few inconsistencies round the coasts of Britain, and the bridled birds

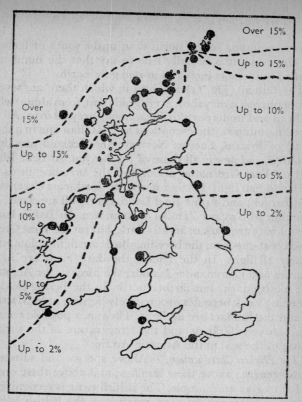

Over 15%

Up to 15%

Up to 10%

Over
15%

Up to 15%

Up to 5%

Up to
10%

Up to 2%

Up to
5%

Up to 2%

Fig. 10. The percentages of the 'bridled' form of the common guillemot at the various British colonies in 1938–9. Derived from H. N. Southern and E. C. R. Reeve.

The gradation of the percentage of bridled birds runs from south to north; the percentage increases more rapidly on the west coast than on the east, though hardly any change takes place over the stretch of the Irish Sea. Probably the proportion of bridled guillemots in the various colonies is slowly changing; and this map represents a stage only in the spread or decrease of the bridled form. The results of Mr Southern's new and very thorough survey in 1949 showed that the proportion of bridled birds was slightly but significantly decreasing at four colonies, and slightly but significantly increasing at one; at no fewer than twenty-five there was no detectable change in ten years. So the change is a very slow one.

are less common in the north than in the south of Iceland. All the same, it is broadly true to say that the number of bridled guillemots increases as one goes north.

The fulmar (Fig. 11) is a bird in which there are several colour phases; it may show every gradation, from almost white on head and underparts to deep, smoky grey all over. All the breeding fulmars (the exceptions are less than one in a thousand) of Britain, Faeroe, Norway, and Iceland are light-coloured, and nearly all those of Jan Mayen. At Bear Island and off East Greenland about half the breeding birds are dark and half light. Farther north, in North-east Greenland, Spitsbergen, and Franz Josef Land, nearly all are dark. The situation in Novaya Zemlya is unknown. In Baffin Island the fulmars are dark or mostly dark, but on the West Greenland coast opposite, the breeding birds are light-coloured, or nearly all light. In the Pacific the dark birds are in the Kuriles and Commander Islands; the birds in the Aleutians are mostly dark; but farther north, in the Bering Sea, the breeding stock becomes progressively lighter; at St Matthew Island the fulmars are all light. There is a possible connexion between darkness and the temperature of the surface-water of the sea, in the Atlantic-Arctic.

The Higher Classification. – Above species and subspecies stand genera; above these families; and above these orders.

Let us give an example. The British twite is recognized as a special geographical race, resident in the British Isles. It has been named *Carduelis flavirostris pipilans* (Latham). Latham was the authority who in 1787 gave the name *pipilans* to this particular bird.

According to the International Rules of Nomenclature, his name has to be put in brackets because he originally attached the name *pipilans* to a different genus (*Fringilla*), from which the bird has now been removed.

The British twite belongs to the subspecies *pipilans* of the species *flavirostris* (to which belongs also the subspecies *flavirostris*, the Continental twite, and others) of the genus *Carduelis* (to which belongs also the species *Carduelis cannabina*, the

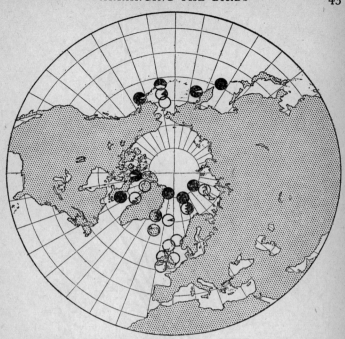

FIG. 11. Breeding distribution of the fulmar. Dark circles: breeding-places occupied by dark fulmars. Open circles: breeding-places occupied wholly or almost wholly by light fulmars. On intermediate circles the approximate proportion of light and dark fulmars is indicated. Circles divided by a jagged border indicate breeding-places where the number of dark and light forms is approximately or probably equal, but from which further information is desirable.

linnet, and others) of the family *Fringillidae* (to which belong also the genera *Coccothraustes* (hawfinches), *Pyrrhula* (bullfinches), and others) of the order *Passeriformes*, or perching birds (to which belong also the families *Corvidae* (crows), *Sturnidae* (starlings) and others).

The family tree in Fig. 12 is an attempt to place the groups of birds on some scale showing their relations in evolution as

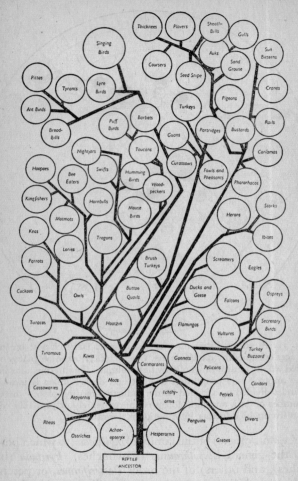

FIG. 12. A Family Tree of Birds, showing what is now suspected of their evolutionary affinities.

far as we can guess at them. The table is a tentative one. It shows bird classification from above downwards rather than from below upwards; it was the latter we discussed when we spoke of species and subspecies. When describing the former it is difficult to know how far down the branches towards the trunk of the tree you should go before you decide that all the smaller branches that have joined your path belong to the same family or order. As with species so it is with these, very largely a matter of man's convenience. The genera, which are very important from the practical point of view, live in a sort of lost world between families and species. They do not often feature in genealogical trees, nor, when the individual animal or bird is handled in the museum, are they discussed as often as species. Their value as biological ideas can be best appreciated by the field worker if he compares several British species belonging to one genus, such as the goldfinch, siskin, redpoll, twite, and linnet, which all belong to the genus *Carduelis*, or the corn-, yellow-, cirl-, and reed-buntings, which all belong to *Emberiza*.

From the ordinary field worker's point of view the important thing is the bird. He is more likely to be interested in the species than in the genus. The one thing essential to him is the power to identify correctly. From whatever point of view he may approach bird watching, he needs skill in identification. This skill is as vital to him if he has the aesthetic approach as if he has a scientific one. If he cannot identify he cannot even enjoy himself with birds. If he begins to watch birds straight off with the help of one of the more old-fashioned text-books, he may get into a bad way about identification, because he will come across the work and the standpoint of the pure museum taxonomist, and the museum man has his own special way of describing birds; thus he can describe every feather but not every pose, and can describe the structure of the voice-box but say nothing about the voice.

So our amateur will find in his books, if they are old ones, detailed descriptions of all that is permanent in a skin, less good descriptions of things like the colour of the eye and legs, and poorish descriptions of any voice or habits. And if after looking at one of these old text-books he goes fussing after chiffchaffs, willow-warblers, and wood-warblers with notions of leg colour or superciliary eye-stripes, he may get himself into a jam.

All the same, identification is the basis of all proper ornithology, and the museum description is the real basis of identification. It is on this that we must build the real description of the bird, and though we may sometimes deprecate it, we must always use it.

Let us trace roughly the course of the average bird-watcher as he discovers the delights and defeats of identification.

1. He becomes interested in birds perhaps because of the influence of somebody else.

2. He cannot make them out, especially when he is alone.

3. So he gets a book with pictures, which may be of two types: (*a*) rather accurate pictures, or (*b*) pictures designed not as pictures but as aids to identification, particularly in regard to colour.

4. Armed with these he improves his knowledge of birds.

5. He begins to get acquainted, not merely with the birds, but also with their habits.

6. With the further weapon of habits he begins to realize the nuances of identification and can speculate about them.

7. Realizing this, he probably widens his circle of ornithological friends.

8. He then becomes capable of identifying birds, not merely by the text-book colours, but also by flight, trickery, song, and all sorts of minor habits of the most subtle kind.

9. He is then in a position to contribute to our knowledge of bird identification.

10. The modern adjuncts to museum descriptions are field-character descriptions, which

11. Themselves are beginning to change museum descriptions.

12. The height of field-character description is reached in the descriptions in the *Handbook*, which are accumulated by field work and note-taking by an ever-widening circle of ornithologists.

Such is the evolution of a man's attitude towards identification. It is not by any means my purpose, in a short and generally introductory book such as this, to try to give a key to the British birds. There are many pocket-books on the

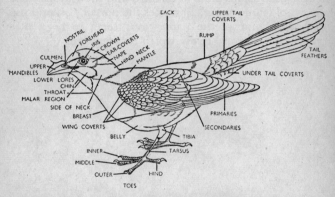

FIG. 13. The externals of a bird, showing the technical names for the different parts of the surface. The names used are those given in the *Handbook of British Birds*; these are carefully designed to have precise meanings to the amateur field worker as well as to the professional ornithologist or museum specialist.

market, some very good, others not so good, designed almost entirely as aids to identification in the field. All I want to say is that you will not get anywhere watching birds without practice, and that you should try to strike a balance between field observation and the use of text-books.

If you are puzzled by a new bird you see, it is not always the best plan to rush straight off to the text-book. You are not likely to have one in your pocket, and the best thing to do is to write accurate notes on every characteristic you can for comparison with the text-book description later. Watch the new bird for as long as you can, note the sort of places it likes, what food it is seeking, its note, its type of flight, and of course its colour, shape, and so on. Any colours or adornments which are clearly of use to the bird for signalling purposes, like white marks on the wings, white rumps or red breasts, are probably equally useful to you in identification. It is very useful to have a fair knowledge of the technical terms for the various parts of the bird and the various tracts and rows of feathers. Fig. 13 names the parts in general use, and though some of the names may seem to be a little cumbersome, they are really worth remembering.

CHAPTER III

The Tools of Bird Watching

THE bird-watcher has to use the following general tools for his job: a library, a note system, optical instruments for viewing (field-glasses, telescope, etc.), and for recording (camera). At one time he often combines two of these general instruments. Besides these he may have occasion to use more specialized instruments for specialized work. These may include traps, rings, stuffed and dead birds, dummy eggs, nestboxes, paint, mirrors, golf-balls, little cakes, tin plates, string, climbing irons, boats, motor-bicycles, balloons, kites, aeroplanes, money, guns, and butter.

The Library. – I do not think it is fair to describe bird watching as an expensive hobby. In a sense it is rather like a stamp collection, in that it can be almost as cheap and certainly as costly as you like. This is abundantly true of one's library. The list of books given below is not meant to be a comprehensive and final one; the books are only suggestions. It is absolutely essential to have at least one good book on the identification of British birds; the rest are a matter of taste and interest.

Among books on the identification and habits of British birds which I can recommend is the late T. A. Coward's *Birds of the British Isles and their Eggs*. This has two editions. The long edition is in three volumes and has coloured pictures (mostly by Thorburn) of almost every bird on the British list. The third volume deals largely with migration and habits rather than with identification, and has pictures of only the very rarest birds. The short edition is really an abridgement of the first two books of the long edition. In this book, which is pocket size, the coloured plates are reduced to quarter-page and really need a lens if they are to be properly used. My own *Bird Recognition*, in this *Pelican* series, will have four volumes, and deals with all the British species that have been recorded more than a hundred times in our islands. It is illustrated in monochrome by *Fish-Hawk* (David Wolfe-Murray). For the novice R. S. R. Fitter's forthcoming field book, to be published by Collins, will identify birds by their colours and shapes, and is to be illustrated in colour by R. A. Richardson.

An expensive[1] book is the *Handbook of British Birds*, edited by the late H. F. Witherby, the late F. C. R. Jourdain, N. F. Ticehurst and the late B. W. Tucker. In my opinion, as soon as the bird-watcher becomes a serious student, he must have a copy of this book or at all events reasonably easy access to it. For instance, it should be his duty to insist that a set is kept in his municipal or county library. The *Handbook* really represents the whole tradition of British ornithology plus the skilful and devoted work of the editors, who were the acknowledged leaders of it. Everything of importance that is known about every bird on the British list is in the book, with its museum and its field descriptions, its distribution at home and abroad, its voice and its display, its habitat and its food, its breeding and its migration. Not only are there pictures in colour of the adult bird but also pictures of the young.

Perhaps the usefulness of the *Handbook* is best realized when the bird-watcher becomes advanced, for if he makes an interesting observation and finds that it is not in the *Handbook*, he can be pretty certain that it is new and worth communicating to the world at large through some local journal or a magazine like *British Birds*.

Such are the text-books which the bird-watcher will find most useful. There are many others on British birds that would help him, some of them remarkably fine, like Kirkman and Jourdain's *British Birds* (London, 1930), and my omission of any names does not mean that I want to underestimate their value. But I do not advise the bird-watcher to

1. The price is high, but the value for money unrivalled.

buy works like Morris's or Wood's *British Birds* or even Howard Saunders's admirable volume unless he is interested from the historical point of view.

As I have already stressed, there is a rapidly tightening link between scientific and amateur bird-watchers. It is therefore worth reading the many general popular works which have been written about birds by scientists. Of these, a few general works might be especially worth looking at. J. Arthur Thomson's *The Biology of Birds* (London, 1923) is quite excellent; and I should like to think that my own *Birds as Animals* (London, 1939) would convince you of the truth of its title. Beebe's *The Bird, its Form and Function* (New York, 1906) is a good American book which has not dated. Nicholson's *How Birds Live* (London, 1927) is excellent.

Books on the general technique of bird watching are really comparatively rare. The best one I can recommend is Nicholson's *The Art of Bird Watching* (London, 1932), which is good because it uses experience and ideas drawn from bird expeditions in parts of the world other than just the British Isles.

Very nearly every important county in Britain has had a book written about its birds, and every year now one or two new books of this kind come out. If you want to find out about published works on birds of your own county the best thing to do is to look in *A Geographical Bibliography of British Ornithology*, by Mullens, Kirke Swann, and Jourdain (London, 1920). This is indexed under counties and contains a list of all known references in bird literature up to 1918. A good list of county and regional avifaunas up to 1945 was published by the British Trust for Ornithology in *Bulletin* 19 (January 1946). A list of local journals was also published in *Bulletin* 18 (April 1945). You will find a county bird book extremely useful, though they are by no means consistent in quality. One of the best things to do is to enter up marginal notes or, if you can afford it, have an interleaved copy made.

There are quite a number of books on separate groups of birds or even on individual birds. Many of them are rather expensive and most of them are highly specialized. It is worth waiting until you develop a keen interest in a particular group before thinking of adding books of this kind to your library. In any case, you will probably find it difficult to get new copies of Millais' *Ducks* or Gurney's *The Gannet*. Of modern books on individual species, it is worth looking at Yeates' *The Life of the Rook* (London, 1934) and Kirkman's *Bird Behaviour* (London, 1937); the latter deals with the black-headed gull. Kirkman used his bird to demonstrate some general principles of bird behaviour. So does David Lack in *The Life of the Robin* (London, 1943), a most stimulating and remarkable book based on long researches. The best modern book on the life-history of a sea-bird is R. M. Lockley's *Shearwaters* (London, 1942). A series of monographs (Collins's *New Naturalist*), newly planned, already (1950) contains *The Yellow Wagtail*, by Stuart Smith, and *The Redstart*, by John Buxton, and books on the greenshank, peregrine, heron, wren, and fulmar can be soon expected.

There are a certain number of books about birds in a special habitat. One of the best is Alexander's *Birds of the Ocean* (New York and London, 1928), which is an essential if you are going on a long sea voyage. E. M. Nicholson's new *Birds and Men* (London, 1950) has a fascinating analysis of the life of birds in towns and on agricultural land, and their adaptation to the ways of man. If you like the history of ornithology, you cannot do better than read the article on Ornithology by Alfred Newton and Sir Peter Chalmers Mitchell in the current edition of the *Encyclopædia Britannica*. If you are interested in the trends of evolution, an inexpensive and useful book is the *Catalogue of Fossil Birds in the British Museum*. Heilman's *Origin of Birds* (London, 1926) is good, and Rothschild's *Extinct Birds* (London, 1907) is interesting but very expensive.

Migration is dealt with in an up-to-date way by A. Landsborough Thomson in his *Bird Migration* (London, 1936, 2nd ed. 1942). If you want to read about bird behaviour, there are many stimulating works, from Frank Finn's *Bird Be-*

haviour (London, 1919) to modern works like Howard's *Terri-tory in Bird Life* (London, 1922, 2nd ed. 1948), Selous's *Realities of Bird Life*(London, 1927), Huxley's *Bird Watching and Bird Behaviour* (London, 1930), Cott's *Adaptive Coloration in Animals* (Cambridge, 1940) and Armstrong's *Bird Display* (London, 1942, new ed. 1947). There has been no comprehensive book on birds' nests since Dixon's book of that title (London, 1902). The best books on bird song are the two volumes by E. M. Nicholson and L. Koch recently published, *Songs of Wild Birds* (London, 1936) and *More Songs of Wild Birds* (London, 1937). These admirable works are accompanied by gramo-phone records made in the field of the songs of most of the common British birds, and they are extremely useful if you want to train yourself for identification by means of song. If you find the subtleties of bird behaviour interesting, it is a good idea to read Fraser Darling's exposition of the life of socially breeding birds, *Bird Flocks and the Breeding Cycle* (Cambridge, 1938).

So far we have not been very helpful to those who have the aesthetic rather than the scientific approach to birds. Such people are, however, admirably catered for. The classic work in English is of course Gilbert White's *Selborne*, though it should be said that its scientific value is as important as its general atmosphere. Lord Grey's *The Charm of Birds* is an-other English classic; it was first published in London in 1927. Most of the late Eliot Howard's work has a high aesthetic and philosophical content. Look at his *A Waterhen's Worlds* (Cambridge, 1940).

Quite a lot of people find the attitude of bird scientists a little difficult, and think that these scientists are surrounding themselves with a barrier of new words and definitions, which is making the truth less easy to get at. I do not believe this is true myself. Though one or two bird scientists are rather pedantic and others rather intolerant, I think nearly all have a very liberal attitude towards ornithology and fully appreciate the work that is being done for birds by those who are not trained as scientists, and the new ideas which such people can bring to the science from their practical experience and intuition. Though I like to think I am a scientist myself, I would not like to feel that scientists could set themselves up as dictators of what has to be done and thought in ornithology. Scientists have no more right to be the bosses than have any of the other kinds of people who have a share in the subject.

There are many different sorts of books about birds which really approach them from the aesthetic side. On the artistic side the work of Peter Scott is particularly worthy of mention. Books like *Morning Flight* (London, 1936) are full of his highly romantic pictures. Sportsmen, by no means a disappearing class in this country, have always had great demands to make on bird artists and writers, and so far as I can see these demands are really aesthetic ones, since most sportsmen, like other artists, prefer atmosphere to achievement. Poets have made both absurd and pointed remarks about birds. This material is best anthologized, and Mary Priestley's *A Book of Birds* (London, 1937) is one of the best of such anthologies.

So much for your library. Of course, my suggestions are only tentative and personal ones. You should not rely on one advice alone. A good beginner's list of bird books was published by the B.T.O. in their *Bulletin* 22 (December 1946). You should plan your library as you think best, and you should be strong-minded about it. It is always best to have some kind of plan, and to get the necessities before the luxuries.

Note-taking. – There are many kinds of systems of taking notes. All of them depend on one basic thing – the field notebook. This should be small and cheap, should have a pencil in an attached tube, or tied by a piece of string. The cover should not be dyed with the sort of colour that runs when it gets wet: black does not usually run. A good idea in a wet part of the country is to keep your notebook in an oilskin tobacco pouch.

What you enter is your affair, as is the sort of shorthand you use. If you are ever undecided about recording something apparently insignificant, decide in favour of putting it in. Number your paragraphs, and start a new sequence and page for each trip. Date the top of this page, and put in a note about the weather. Always remember what county you are in, or better still, the vice-county. The British Trust for Ornithology (which is the chief field-work organization in Britain) has standardized the vice-county system for field work. The vice-counties of Britain, originally designed for botanical work, are shown in Fig. 14. The vice-county map is very useful, as these divisions are of very manageable areas and represent a good compromise between nature- and man-politics.

Broad geographical distribution of, say, a breeding bird can be very neatly shown if you simply block in the vice-counties from which there are positive records. The results of doing so are well shown in my *Bird Recognition* (Pelican), whose maps are compiled by my friend W. B. Alexander.

If you are accompanied in your walks by ornithological friends, put down their initials at the top of the page, so that you can later refer to them for any necessary corroboration of fact. Use the Continental time-system, i.e. 15.30 for 3.30 p.m. Use ♂ for male and ♀ for female. Write down as many details of bird numbers as you can. If actual numbers cannot be estimated, it is a good idea to put:

Order 1. Under 10 birds.
Order 2. Under 100.
Order 3. Under 1,000, and so on.

Never record a bird as definitely breeding unless there is positive evidence for the existence of eggs or young.

Never record an identification you are doubtful of without writing down why you are doubtful. Record all hearsay evidence that you meet with, but record it critically.

Write up the results of your field-work out of your field notebook into your permanent record *at the earliest possible moment*. If you leave things too long you may forget your own abbreviations. Cross out the notes in your field notebook as

ENGLAND AND WALES

PENINSULA
1 West Cornwall with Scilly
2 East Cornwall
3 South Devon
4 North Devon
5 South Somerset
6 North Somerset

CHANNEL
7 North Wilts
8 South Wilts
9 Dorset
10 Isle of Wight
11 Hants South
12 Hants North
13 West Sussex
14 East Sussex

THAMES
15 East Kent
16 West Kent
17 Surrey
18 South Essex
19 North Essex
20 Herts
21 Middlesex

22 Berks
23 Oxford
24 Bucks

ANGLIA
25 East Suffolk
26 West Suffolk
27 East Norfolk
28 West Norfolk
29 Cambridge
30 Bedford and detached part of Hunts
31 Hunts
32 Northampton

SEVERN
33 East Gloucester
34 West Gloucester
35 Monmouth
36 Hereford
37 Worcester
38 Warwick
39 Stafford and Dudley
40 Shropshire

SOUTH WALES
41 Glamorgan
42 Brecon
43 Radnor
44 Carmarthen
45 Pembroke
46 Cardigan

NORTH WALES
47 Montgomery
48 Merioneth
49 Carnarvon
50 Denbigh and parts of Flint
51 Flint
52 Anglesey

TRENT
53 South Lincoln
54 North Lincoln
55 Leicester with Rutland
56 Nottingham
57 Derby

MERSEY
58 Cheshire
59 South Lancashire
60 Mid Lancashire

HUMBER
61 South-east York
62 North-east York
63 South-west York
64 Mid-west York
65 North-west York

TYNE
66 Durham
67 Northumberland South
68 Cheviotland, or Northumberland North

LAKES
69 Westmorland with North Lancashire
70 Cumberland
71 Isle of Man

SCOTLAND

W. LOWLANDS
72 Dumfries
73 Kirkcudbright
74 Wigtown
75 Ayr
76 Renfrew
77 Lanark and E. Dumbarton

E. LOWLANDS
78 Peebles
79 Selkirk
80 Roxburgh
81 Berwick
82 East Lothian

83 Midlothian
84 West Lothian

E. HIGHLANDS
85 Fife with Kinross
86 Stirling
87 South Perth with Clackmannan, and parts of Stirling
88 Mid Perth
89 North Perth
90 Angus or Forfar
91 Kincardine
92 South Aberdeen
93 North Aberdeen
94 Banff

95 Moray or Elgin
96 Easterness (East Inverness with Nairn)

W. HIGHLANDS
97 Westerness (West Inverness with North Argyll)
98 Argyll (Main)
99 Dumbarton (West)
100 Clyde Isles
101 Cantire
102 South Ebudes (Islay, etc.) and Scarba

103 Mid Ebudes (Mull, etc.)
104 North Ebudes (Skye, etc.)

N. HIGHLANDS
105 West Ross
106 East Ross
107 East Sutherland
108 West Sutherland
109 Caithness

NORTH ISLES
110 Outer Hebrides
111 Orkney
112 Shetland

IRELAND

(1) 113 South Kerry
(2) 114 North Kerry
(3) 115 West Cork
(4) 116 Mid Cork
(5) 117 East Cork
(6) 118 Waterford
(7) 119 South Tipperary
(8) 120 Limerick
(9) 121 Clare with Aran Isles
(10) 122 North Tipperary

(11) 123 Kilkenny
(12) 124 Wexford
(13) 125 Carlow
(14) 126 Leix
(15) 127 South-east Galway
(16) 128 West Galway
(17) 29 North-east Galway
(18) 130 Offaly
(19) 131 Kildare
(20) 132 Wicklow

(21) 133 Dublin
(22) 134 Meath
(23) 135 Westmeath
(24) 136 Longford
(25) 137 Roscommon
(26) 138 East Mayo
(27) 139 West Mayo
(28) 140 Sligo
(29) 141 Leitrim
(30) 142 Cavan
(31) 143 Louth
(32) 144 Monaghan
(33) 145 Fermanagh

(34) 146 East Donegal
(35) 147 West Donegal
(36) 148 Tyrone
(37) 149 Armagh
(38) 150 Down
(39) 151 Antrim
(40) 152 Derry

KEY TO FIG. 14, OPPOSITE.

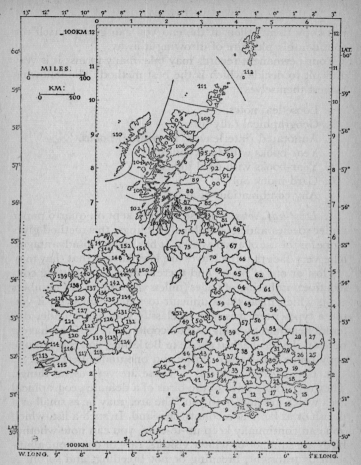

The political divisions of Britain, i.e. the simple counties, are of all sorts of different shapes and sizes, and differ greatly in biological importance. The idea of the vice-county system is to divide Britain into areas of roughly equal importance, without sacrificing the county system. A glance at the map will show the sort of compromise that has been reached.

(Reproduced by permission of Messrs Collins, publishers of the *New Naturalist*.)

FIG. 14. The Counties and Vice-Counties of the British Isles.

KEY TO MAP IS OPPOSITE

you copy them; then at the end you can give yourself the considerable pleasure of throwing it away.

Your permanent records may take many forms; it is very difficult to decide which is the best method. The following suggest themselves:

1. Loose-leaf notes.
2. Geographical tally-lists.
3. Annotated (interleaved, if possible) books.
4. Year-books with back index.
5. Year-books with card index.
6. Card index only.
7. Any combinations of above.

1. *Loose-leaf Notes.* – These are best kept on quarto paper under species and geographical headings; this method gives scope for the use of a typewriter. It has several disadvantages. The very fact that the leaves are loose means that they may get lost or out of place, and there is no way of getting consecutive numbering of pages (unless you add 'a', 'b' numbers etc.), which means it is difficult to index. However, if you use a typewriter, it is easy to classify your notes under different headings by using carbon copies; that is, if you have a note on the red-backed shrike in Berkshire, you can put one copy under red-backed shrike and one under Berkshire.

2. *Geographical Tally-lists.* – These are very useful things. They are simply a list of the birds of a definite geographical area which are known to you. The area may be as small as a parish or as large as a county or island. In such a list, which you can continually keep up-to-date, you can note whether birds are summer or winter visitors, rare or common, abundant or transient, breeding or not breeding, and so on. A skeleton list of this sort is published by the British Trust for Ornithology, with the hope that members and friends of the Trust can fill in the form and put a copy of it on deposit at Oxford for permanent record. This form can be obtained from the Librarian of the Edward Grey Institute, 91 Banbury Road, Oxford, and he will also give you details of cost. A

similar form is Sir Hugh S. Gladstone's *An Ornithologist's Field Notebook* (London, Truslove & Hanson), and small convenient pocket tally-forms are also published by the London Natural History Society and the Birmingham Bird Club.

The keeping of a tally-list is a very good exercise in itself and it is also a great stimulus to a beginner or to a young ornithologist. Most people start ornithology by making a list of all the birds they have seen, and if this is linked up with the study of an area, the result may be of permanent value to science. But the tally system does produce certain peculiar results. It tends rather to focus people's attention on the variety of birds they can see, and among a certain section of ornithologists it induces useless rivalry. I think the proper attitude towards tally hunting is that it is a useful thing to do provided it does not become too much of an end in itself. It is of very much greater importance to get a thorough knowledge of the distribution and habits of the common birds in your area than to go chasing rarities.

3. *Annotated (Interleaved, if possible) Books.* – Many ornithologists, particularly those who can afford a fairly extensive library, make marginal comments in their books, and sometimes go to the trouble of having them specially interleaved. That great ornithologist, the late F. C. R. Jourdain, used this method, and his annotated books have many times their original value to science because of the importance of his notes. Perhaps the senior generation of bird-watchers use this method more than the younger people. It is probable that the card-index system answers just as well, and it is certainly difficult to know whether posterity is best served by having notes in books already published or by having them under a subheading of their own; and very often a marginal note is lost if the book is disposed of.

4. *Year-books, with Index at the End.* – Most of those who have the notebook in a highly developed form are divided between the merits of year-books and the pure card-index system. A year-book is a strongly bound exercise book, kept in the form of a diary, from extracts from the field notebook.

Most people use large margins or only one side of the paper, and underline places in red and species in black, or they may put the place in capital letters in the margin opposite the relevant note. Blank pages leave scope for pasting in photographs, press cuttings, and even portions of feathers. At the end of the year the remaining space in the book is used to make an index of page references under headings of species and geography and various subjects. A list of such subjects may be as follows:

Weights	Nests
Measurements	Eggs
Temperature	Incubation
Length of life	Hatching
Food	Fledging
Ecology	Flight
Population	Identification
Migration	Sex difference
Rings used	Recognition
Mortality	Use of colours
Predators	Variation
Prey	Inheritance
Behaviour	Diseases
Display (with courtship)	Growth
Flocking	Development
Voice	Anatomy
Song	Moulting
Alarm notes	Miscellaneous

5. *Year-book, with a Card Index.* – For the index in Section 4, which can be written up only at the end of the year and which has to be written up every year, some people substitute a card index referring to the year-book. This can be kept up concurrently with the entries and need not be started again every year if the volume numbers of the year-books are given. In many ways this is a very good system, since quite small cards, 5 in. by 3, can be used, and the index is infinitely expansible. The disadvantage of the system is that

it is a duplicate one and that you cannot carry a card index about with you.

6. *A Card Index Alone.* – There is very little to choose in efficiency between this method and the previous one. I use this method myself, but I often wonder whether it is really better than the year-book system. If a six-by-four card, which is the most manageable size, is used, you cannot always get all the notes you want about an observation on to the card and have to overflow on to another. Many people have different card-index systems; mine is roughly as follows:

I have three simultaneous card indexes, each referring across to the others. They are (*a*) a subject-index, (*b*) an author-index, and (*c*) a species-index.

The subject-index runs in an order which I have more or less worked out for myself. It is not possible to lay down a universal system, because your own tastes will naturally alter its scope and balance.

The author-index consists of references to books and scientific papers arranged alphabetically under the names of authors. When references to more than one work by an author appear, they are in order of date.

The species-index consists of cards for all the birds, arranged in the order set out by the Zoological Society in the *Vertebrate List (Birds)*. The standard *Check-list of Birds of the World*, by J. L. Peters (Cambridge, Massachusetts), started to come out in 1931, but is as yet only half completed. My interests lie with birds from all parts of the world, and a student of British ornithology might be better advised to arrange his birds in the order of the *Handbook* or of the British Museum list. If he uses the *Handbook* he will find that all the birds, both species and subspecies, on the British list have numbers; this is, of course, very useful. At the moment workers on each side of the Atlantic classify the birds somewhat differently; it is to be hoped that a universal system will some day be adopted.

In my particular index, cards of ten different colours may be found under the name of one species. These cards refer to

(1) name, authority for that name, references to pictures in the literature, and broad geographical distribution; (2) all matters dealing with bird's zoological position and classification; (3) all measurements, including weights, linear measurements, temperature, pulse rate, respiration rate, and length of life; (4) food; (5) ecology, including all matters of restricted or specialized geographical distribution, relations with predators or prey, population, changes of range, and migration; (6) behaviour in general; (7) reproduction, including all details of nest construction and the management of eggs and young which cannot be included under behaviour; (8) variation, heredity, embryology, and development; (9) disease and death; and (10) anatomy, physiology, and miscellaneous items.

On any of these cards information can be found under the appropriate heading about the particular species whose name and number are on the card. This information may consist of a note of my own copied straight out of my own field notebook, or it may consist of a cross-reference to something already published, under author and date. If so, there will be a short résumé of the relative material on the coloured card underneath the author and date. If I want to look up more about it I go over to the author-index and take out the card for that author and date. This gives me details of where the book or paper can be found (for instance, whether it is in my own library or whether it is to be found in some library to which I may have access, and if so, on what shelf). I recommend this system very warmly to anybody working in a large public library like the British Museum or one of the university libraries, since you will save a great deal of time by having your press mark ready. Ornithologists should be encouraged to save each other time by exchanging press marks if the opportunity should arise.

Field Glasses and Telescopes. – There is only one moderately expensive tool which need stand in the way of anyone's evolution from a keen bird-watcher to a first-class ornithologist – this is, a good pair of field glasses. Most of the glasses on the market vary from cheap, small, and heavy pieces with a low magnification of × 4 or × 5 and a tiny field, to weapons like the most modern ultra-lightweight Ross or Zeiss glass, with a magnification of about × 9 and an extra-wide field of vision.

Between the guinea or two that you waste on the first and the 25 to 35 guineas that you would not grudge for the last (if you had it), there is a wide range of good instruments which should suit you well if you have gone the right way about your buying. Unless you can afford big glasses with an extra-wide field, it is not worth getting any with a magnification of more than × 7 or × 8, and I do not see anything wrong with a pair at × 6. As your magnification goes up, your wideness of field decreases unless it is compensated for by increased size of the binoculars.

Heaviness of glasses or smallness of field are both factors which prevent you from following a quickly moving bird. If your field is too small you may have great difficulty in picking up a bird you have previously sighted with the naked eye, and hunting a field by moving your glasses up and down is a tiresome strain on the eyes.

So when you buy your glasses, try to get a pair which is light, which has a magnification of × 9 or under, which has a wide field, and which is free from chromatic aberrations (for it is no use seeing a rainbow round every brightly illuminated object in the field of view). On some types of binoculars lightness is secured by the abolition of a central focusing screw and the substitution of independent eyepiece focusing. This means that the observer has to balance his binoculars on his palms and work each eye-piece with thumb and forefinger, a job which takes both hands and a fair amount of practice. Although it makes the glasses heavier and a little more expensive, I would plump for the central focusing screw, provided there is a device on one of the eyepieces by which an adjustment can be made for personal differences between your two eyes.

Carrying your glasses is quite an art. A leather case is absolutely essential for long journeys, but by no means so on every bird walk. On such walks I know of two methods of protecting your glasses without carrying a cumbersome case. My father used a chamois-leather bag which slipped over the glasses and tied in a bow at the top. E. M. Nicholson has a very neat light leather gadget which slides up and down the straps and which fits just over the eyepieces to protect them when it is raining. I imagine bird-watchers must have worked out many such devices.

Telescopes are useful things, but are probably not worth the money unless you have special problems to solve. I have a very nice light one, but lately I have found myself carrying it many miles without using it. It is most useful for watching birds on an open nest from a distance or a flock of geese in a meadow or a group of waders some hundreds of yards away over soft mud at low tide; in fact, under circumstances when birds are very shy or inaccessible. But practically all birds met with tolerate man within range of an ordinary pair of field glasses. A light telescope with a magnification of perhaps 25 or 30 diameters is rather difficult to manage without a stand; the field is so small that the slightest movement may lose you your bird. Though you can use a neighbour's shoulder or lie down and balance the telescope on your feet, most people find that a proper telescope stand becomes necessary. These are usually made of wood, though my father and I got very useful and light ones home-made, out of metal tubing. The kind with telescopic legs, like a very big camera stand, are very expensive. If one has to generalize about telescopes, it might be safe to say that the people who find most use for them are those who do a lot of their bird watching from a car, out of which they can hop to pull out their telescope and stand. For mountain work or any stiff cross-country work I would discard the telescope altogether, as you want to be as unencumbered with gear as possible. Some people carry a simple stick whose length is the same as the distance of their mouth from the ground, and rest their field-glasses on top of it; but this is no good for telescopes.

The Camera. – This book is not addressed to budding bird

photographers. This science or art is catered for very adequately by organizations like the Zoological Photographic Club and by books like Hosking and Newberry's *Art of Bird Photography* (London, 1944), or George Yeates's *Bird Photography* (London, 1946). It is difficult to say whether the real bird photographer is concerned more with birds than with photography. In any case, the result is an art which is by no means pure ornithology, though it is a very stimulating, interesting, and patient art. My business is not to tell you how to become a bird photographer – not that I could if I tried – but to suggest one or two ways of using a camera as an accessory method of making bird records and taking bird notes.

The camera I use myself is a Leica and is to my mind ideally suited to this particular purpose. I have used it from a hide with a telephoto attachment, on mountains, over cliffs, up trees, indoors, and on land and sea out of doors. With a miniature camera you can take permanent records of bird behaviour, particularly at the nest; you can take a whole battery of exposures one after another and get an almost continuous record of a bit of behaviour; you can have a permanent record of a field experiment (for instance, the reactions of birds towards a stuffed bird) more quickly and accurately than you could ever do with a notebook; you can take pictures of your friends watching birds and birds watching your friends; you can take pictures of general habitats as well as close-ups of individual birds. All these things, of course, you can do with other types of camera, and many of these cameras, especially large reflex cameras, give better and more artistic results. But if your attitude, like mine, towards a camera is that it should be a tool for general purposes, you cannot do better than use a miniature one.

With a miniature you can afford to take large numbers of pictures and to document everything you have a mind to. With a miniature you are unlikely to win prizes at a photographic exhibition, but more than likely to increase our knowledge of birds and your own satisfaction in them.

CHAPTER IV

Migration

❦

We must not, I think, deny migration in general: because migration certainly does subsist in certain places, as my brother in Andalusia has fully informed me. Of the motions of these birds he has ocular demonstration.
GILBERT WHITE, 12 February 1772

❦

IN practically every group of animals, wherever these animals live, there can be found members which indulge in orderly mass movements. These movements may happen every few hours, every day, every few days, every month, every year, or every few years. Animals do not, as far as I know, have regular movements of the order of every week, because the week is a human invention which bears no relation to natural phenomena like the daily turning of the earth, the transit of the moon, or the movement of the earth round the sun.

The lives of birds are based on an annual rhythm; the only other important rhythm we can find in their lives is the periodic fluctuation in the numbers of some species which produce a high population every fourth to fifteenth year or so, and a low population at some period between the high peaks. Such movements as are connected with these yearly or several-yearly rhythms we call migrations.

The annual rhythm may produce several sorts of migration. It may consist of mere local movements or dispersals in no particular direction. It may consist of the typical south-to-north movement that makes it possible for us to label such a large proportion of the birds on the British list as summer or winter visitors. This south-to-north movement may even involve crossing the Equator. In the tropics, however, there may be orderly seasonal movements, but these may not be connected at all with latitude. The birds move every year at the best time to avoid the dry season, and these movements every year may be quite irregular in their nature and yet orderly because they always happen at the same season. When we discuss the special case of rhythms extending over not one but several years, we find ourselves dealing with the remarkable cases of periodic irruptions, extensions of birds when their population is high into geographical areas not usually within their natural range.

1. *Local Movements.* – Many kinds of birds are as a rule described as residents, because they are never known to move very far from where they are hatched. Yet every year these birds move about locally, and sometimes this movement is more or less orderly. Some birds move up and down mountains. Some form flocks in winter and leave their breeding territories to roam together about the hedges and fields of the countryside in search of food. The yellowhammer and chaffinch are typical British birds which never appear to migrate in the true sense, but which flock up and roam about open country. Similarly tits and goldcrests, highly resident birds, flock up in autumn and go on foraging parties for the winter.

There seems to be scope for a fair amount of individual variation, since some individuals of normally resident birds may carry out true migrations. For instance, nearly all British robins are resident; except for a substantial minority of the females, which migrate. Scarcely any males migrate; they all stay at home and mark out their spring territories.

Local movements start rather earlier than most people imagine, since even before the parent birds have finished with their last brood, the offspring of the first broods may have been driven away and may have scattered through the country. These young birds often join with the flocks and quite often roost in company. The early autumn movements seem quite highly organized and deserve a good deal more study than they have had so far.

2. *Dispersals.* – The movements of many birds are governed by the necessity for breeding in the summer season, but by no particular necessity to move southwards in winter. Yet the movements of these birds are too large and too definite to be called local movements. This especially applies to sea-birds, which may disperse for many hundreds and some-times even thousands of miles from their breeding haunts. Just before writing this chapter I heard of two puffins marked on St Kilda, the most westerly of the Scottish islands, which had crossed the Atlantic. They must have done this as much by swimming as by flying. The distance was over 2,000 miles. Just before I revised this book for the present edition, two fulmars which I had marked myself on St Kilda in July 1948 were recovered near Newfoundland in the following November and June.

Sea-birds have the whole of the ocean to feed in, but some choose only a limited number of sites to nest on. Dispersal is therefore very wide, and gannets, which breed only in thirteen colonies in Britain and twenty-nine in the world, spread out all over the Continental shelf, where the water is not too deep, in the winter. The direction of their spread may be north, west, east, or south; where the bird goes seems to depend largely on its individual inclinations. This is not quite true of young gannets, which seem, during the three years before they can breed, to indulge in true north-and-south migration, reaching the coast of North-west Africa.

Kittiwakes, fulmars and other petrels, and most of the auk tribe, are oceanic birds in winter. Their spread from their breeding haunts may indeed lead them south in the arctic regions, where ice occupies the open sea, but where there is no ice their movements seem to be in the direction of most food or in no particular direction. Indeed, as far as I can detect, the general distribution of the oceanic birds in the North Atlantic is primarily determined by food-supply, and all species can tolerate a very considerable range of temperature and weather in their search for food. If some species, like the little auk, are found farther south in mid-ocean in winter than summer it is not, I believe, because they are migratory so much as because their basic food-supply shifts.

3. *Typical Migration.* – Most of our summer visitors and nearly all our winter visitors in Britain are typical migrants, that is, they are birds which have two distinct ranges, though in many cases these ranges may overlap. Between these ranges they have an orderly movement in spring and autumn. When they are making this movement we may say they are on passage. There are certain birds, like bluethroats and Greenland wheatears, whose two ranges do not overlap, the summer range being north and the winter range south of Britain. These birds visit us only on passage and appear in the British list as passage migrants.

Britain is situated in a remarkably good position from the ornithologist's point of view. It lies athwart the main migration stream along the west coast of Europe. Every autumn the main streams of migrants pour in four great passages along its shores. The greatest of these passages is down the east coast from Shetland to Kent. Birds from Scandinavia cross the North Sea to form this stream; when it reaches Kent it is joined by the second stream, which has come down the western coast of Denmark, Germany, and the Low Countries, passing the famous migratory observation station of Heligoland. The joint streams then pass along the south coasts of Britain and the north coasts of France.

The third stream is composed of some of the Shetland and Orkney birds and of some from the Faeroes and Iceland. It runs westwards across the top of Scotland and down the west coast as far as Northern Ireland, where it continues down the coasts of the Irish Sea, having given off the fourth stream, which runs down the west coast of Ireland. Both these streams meet each other and the remnants of streams one and two in the south, and may continue down the west coast of France. Fig 15 shows the main nature of these streams.

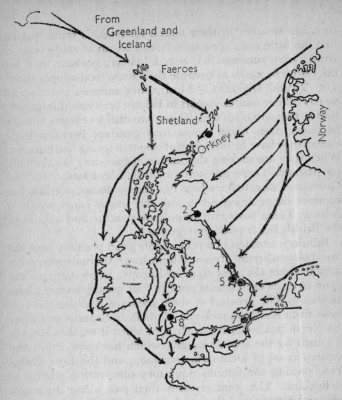

Fig. 15. The main paths of the autumn migration along the shores of Britain, and the positions of some of the more important field stations and observation points.

These are:
  (1) The Fair Isle, Shetland.
  (2) The Isle of May, Fife.
  (3) Bamburgh, Northumberland.
  (4) Spurn Head, Yorkshire.
  (5) Gibraltar Point, Lincolnshire.
  (6) Scolt Head, Blakeney Point, and Cley Marshes, Norfolk.
  (7) Dungeness, Kent.
  (8) Lundy, Devon.
  (9) Skokholm, Pembrokeshire.

(Adapted from Landsborough Thomson.)

4. *Trans-equatorial Migration.* – In some cases the two ranges between which birds migrate are on opposite sides of the Equator. Arctic terns migrate from arctic to antarctic, breeding only in the northern hemisphere. Many sea-birds cross the Equator every year to their winter quarters in the southern seas; some, like the great shearwater, breed in the south and spend a non-breeding season in the summer north of the Equator.

On the whole there are very few birds which breed in the southern hemisphere and winter in the north. If you look at a map you will see why this is so. The northern temperate regions have the huge areas of North America, Europe, Asia, and North Africa; the tropics have the middle-sized areas of tropical South America, Africa, and the isles of the East; while the temperate southern hemisphere has only the southern part of South America, the tip of South Africa and South Australia and New Zealand, which are tiny areas by comparison with the vast temperate areas of the north.

Among land birds there are plenty of examples, like the swallow, cuckoo, and stork, which breed in the northern hemisphere and winter in the south. But there is not one single example known to science of a *land* bird which breeds in the southern hemisphere and winters in the northern.

5. *Movements at Several-yearly Intervals.* – It is becoming clear that a far larger number of birds than was at first suspected undergo more or less regular changes in population. When a peak of population is reached, it very often means that the pressure of numbers causes a big movement which it seems reasonable enough to call a migration. Some authorities describe these as irregular migrations or irruptions. Perhaps the latter term is better, since one of the characteristics of these mass movements is that the interval between them is often very much the same number of years, very often between nine and twelve, and almost always between five and sixteen.

In Britain irruptions of the crossbill arrive at quite short intervals (every three to ten years), and in some years these

have been so large that a remnant has actually stayed to
breed (see Fig. 16). The permanent colony of crossbills in
Norfolk was established after one of these irruptions. Pallas's

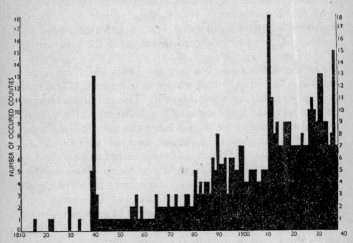

FIG. 16. The crossbill in Britain over a period of 130 years.

This graph or chart shows what the Continental crossbill has been doing in Britain
during this period. The horizontal axis runs in years and the vertical axis represents the
number of counties that in each year are known or can be reasonably supposed to have
crossbills breeding in them. The main conclusions to be drawn from the chart are, first,
that a large number of irruptions of crossbills, of various degrees of importance, have
happened in the period under review. Second, that in the years 1838–40, 1899, 1910,
1927–30, and 1936, these irruptions have been very large. Third, after big irruptions, the
birds have stayed permanently to breed in one or more counties.

To-day, besides the permanently breeding crossbills in Norfolk and Suffolk, there are
several stocks that seem to have definitely settled in parts of Ireland. There are now
probably never fewer than six or seven counties with crossbills breeding in them every
year. But besides this permanent breeding stock, crossbills still irrupt from outside every
three to ten years or so. The last influx but one, which took place mainly in 1927–28, was
the biggest for many years, though the score in 'new counties' was not so great as in
some other years.

sand-grouse used to appear in Britain in large numbers about
every twelve years, but has not done so now for forty or fifty
years. Rose-coloured starlings seem to have a cycle of nine
years or so, and at the peak of this often irrupt from Asia to

Europe and breed for a year or two. In these peak years birds quite often reach Britain.

Migration is an expensive process, since no bird is a perfect migrant. Nobody fully understands how migrants can find their way (some, like the Pacific golden plover, regularly cross 2,000 miles of ocean without a landmark), or how a young bird which leaves for a new country some time after its parents have left develops any sense of what to do. But in spite of this wonderful adaptation to making journeys without maps, birds quite often make mistakes, and quite often fail to complete their journeys through exhaustion. Authorities have recorded many instances of birds being lost in fogs, and failing to pick up such coastal landmarks as they are known to use; and there are innumerable cases of high mortality after storms, of birds driven miles out of their normal routes to places where they are extremely rare, and of exhausted birds seeking shelter on ships at sea.

The enormous difference in the number of birds coming up in the spring migration and those going down in autumn must be entirely due to losses in winter and on migration, and there can be no doubt that migration takes a terrible toll of life. Clearly, then, its high cost in bird material must be offset by some very high biological advantage.

The mechanism of migration is extremely complicated and by no means fully understood. Many writers call it mysterious, and they have a good deal of justfication for this; though it must be said that propaganda about mysterious forces often puts a very effective brake on the scientific urge towards discovery. The origin of the habit of migration, looked at in the perspective of evolution, must be a story of trends towards orderly movement built up and improved by natural selection. These trends themselves have taken place in a changing environment; ice caps and other geographical barriers have come and gone, and on this pattern of historical movement the migratory habit has been built up.

The biological advantages (which I have suggested must exist to justify part at least of the migratory habit) fall under the heads of climate (warmth and humidity), light, food-supply, and habitat. Food-supply is probably the most important.

A bird that moves north in the summer to breed, by doing so continues to live at the temperature to which it is suited. A bird that breeds in the high arctic has a fairly rich insect life newly released by the retreating snow on which to feed its nestlings, and it has twenty-four hours of daylight in which to catch these insects. In the tropics birds move in an ordinary migration to breed, not necessarily in cooler or lighter parts of the world, but in parts where the fruitful wet season persists and where the barren dry season can be avoided.

By migrating many birds must find, not only a climate more suited to them, but also a place to build their nests. For the distribution of birds depends to a certain extent on the availability of nest sites.

Technique of Study. – Since the very earliest times – times when the extent and nature of birds' migration was scarcely realized – observers of the seasons have noticed the arrival and song of birds. Most people find that in their own personal history as bird-watchers, arrival and to a lesser extent departure dates have featured largely in their notes. To-day the arrivals of migrants are widely recorded in every part of the civilized world, and in Britain I know of no natural history society which does not pay some form of organized attention to them. From what we know of our dates we can construct a very good picture of the advancing path of spring migration through Europe. Mr H. N. Southern has done this very thoroughly for the swallow, the willow-warbler, and the redstart, and I am grateful to him for permission to reproduce one of his excellent maps as Fig. 17.

Though the autumn migration is a bigger affair than that of spring, our knowledge of departure dates is not nearly as good as that of arrival dates, and the subject deserves a good deal more careful attention than it gets at present. It is safe to say that the observation of individual birds on migration and the noting of dates is still our first line of attack on the secrets of migration in general. The offensive in this particular field, the observational front as it were, was taken by Dr W. Eagle Clarke, working with a committee of the British Association for the Advancement of Science, over sixty years ago. He organized the collection of observations on birds migrating from as many reliable individuals as he knew and

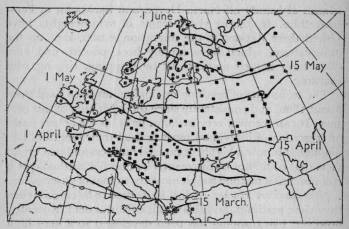

FIG. 17. The rate of spread of swallows over Europe in spring.

Swallows arrive in Europe in spring over a broad front, and reach the very south a full two and a half months before they reach the very north. There is a tendency for the movement to be more rapid along the west coast and less rapid in the mountain areas. The thick black lines show the position of the main advancing front on the dates marked and the square dots are the places from which the necessary observations were taken. (After H. N. Southern.)

from as many points along the coast of Britain as was possible. Perhaps his most valuable contribution was his development of a questionnaire technique and the use of all reliable lighthouse keepers and lightship crews that he could find.

This sort of work was continued in Ireland by Barrington and has now been extended by many individuals and societies to practically the whole coast of Britain. More recently key points have been developed as organized field stations or temporary observation points. Many of these are marked in Fig. 15. They include Fair Isle in Shetland, the Isle of May in the Firth of Forth, Bamburgh and Holy Island off the Northumberland coast, Spurn Head in Yorkshire, Gibraltar Point in Lincolnshire, the stretch of the Norfolk coast from Scolt Head to Cley Marshes, and on the west coast the islands of Lundy off North Devon and Skokholm in Pembrokeshire. In recent years teams of ornithologists have visited these stations in the migration season, kept up the observations, which have usually been pooled, and very often have helped matters out by trapping, ringing, and marking.

The technique of marking or ringing has been used since about 1890, and consistently since the beginning of this century. It can throw a very brilliant light on the actual paths of migration of individual birds and on their length of life. Though it is now a highly organized business and can reveal facts that could not be discovered by another method, yet it still remains a method supplementary to that of observation.

How Birds are Marked. – Most birds used to be ringed when they were nestlings, and this is still a fruitful field. Recently, however, live trapping has been developed to a very high pitch of excellence, and at present about half the 50,000 birds ringed annually in Britain are done when adult and half when young. Many kinds of live traps are described in P. A. D. Hollom's *Trapping Methods for Bird Ringers*, the British Trust for Ornithology's Field Guide No. 1.

(*a*) *The Sieve-trap.* – This is a development of the old method of propping a sieve up on a stick and pulling the stick away by a piece of string when the birds are underneath. A modern sieve- or drop-trap is rectangular and usually made with wooden sides. The top is covered with either wire-netting or, better, string netting, and at one corner there is a door opposite which a catch-up can be placed. A catch-up is merely a box with netting at one end and a drop door at the other into which the bird can be driven from the main trap. Once the bird is in the catch-up you can reach and grasp it gently.

(b) *The Clap-net*. – This may be double or single, and lies flat on the ground. It consists of two poles hinged to pegs driven flush in the ground. Between them a net is stretched; to the loose end of one of the poles a pull-string is attached. When this is drawn it lifts the poles off the ground and drops them and the net over on to the other side, so that the net falls on top of the birds, feeding along a line of baits. It is rather difficult to get birds out of a clap-net because they struggle and get entangled. The best way is to place your fingers in the meshes of the net round the bird and then spread them out so that each mesh is stretched wide.

(c) *Funnel-traps*. – These may be of all kinds, from large ones for rooks to small sparrow traps. The principle is quite simple. The trap is baited and entered by a funnel. This tapers so that once the bird is in it cannot find its way out. Big rook traps sometimes have a vertical funnel which may be directly entered from flight.

(d) *The House-trap*. – This is a recent development much used in Oxford and Cambridge. It consists of a big box of wire netting, in which a man can stand, which is entered by double wire doors. To receive the bird these doors are merely left slightly ajar. As they open in opposite directions the birds do not find their way out.

(e) *The Heligoland Trap*. – This type is in common use in migration stations. It consists of an immense tunnel tapering and curving gradually to a reception chamber into which the birds fly or are driven. At the mouth of the tunnel, which may be 30 feet or more across, a lot of cover is planted for the birds, and they can sometimes be very easily driven in without heavy baiting.

(f) A useful trap for gardens is the *Potter trap*. This consists of a series of baited compartments. When the birds go in they tread on a platform which drops a door behind them.

(g) *Bat-fowling Nets*. – These are long nets held on bamboos; at the top the bamboos are hinged and the pair can be

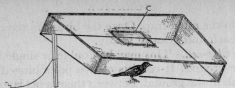

A. Simple sieve-trap (fall-trap).

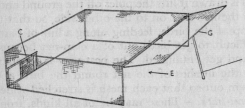

B. Modified sieve-trap (drop-trap).

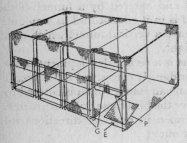

C. Potter trap.

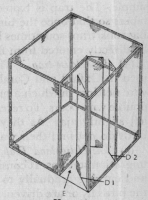

E. House-trap.

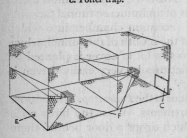

D. Funnel-trap.

C. *Door for catching-up.*
D. *Doors for house-trap, which acts as funnel, preventing escape of birds, and at same time allows entry of human beings to clear traps.*
E. *Entrance for birds.*
F. *Funnel.*
G. *Gate which drops and captures birds.*
P. *Platform which drops gate when actuated by weight of bird.*

FIG. 18. Some types of bird-traps (not to scale).

clapped together so as to drop the net round a flying bird. They are very useful for catching birds at roosts at night, but not much good during the daytime.

(h) *Kingfisher-nets*. – These consist really of two nets, one with a very large mesh and the other with a small one. When a bird flies into one of these nets it pulls the small-mesh net through one of the meshes of the larger-meshed net, thus making a pocket of net from which it cannot escape. These used often to be stretched under bridges to catch kingfishers for the plume trade. Nowadays they are put to better use by students of migration. Of course they are applicable to many sorts of birds besides kingfishers, and can be stretched from tree to tree just as easily as under a bridge.

(i) Not many *duck-decoys* exist in working order to-day, but one or two, like that at Orielton, and that of the Severn Wildfowl Trust in Gloucestershire, are still kept up in order to mark ducks.

Trapping is great fun in its own right, and many people with small farms or country houses are able to mark well over 1,000 birds a year. Naturally nobody should use traps of the type which the birds themselves actuate or which they may enter at any time of the day, unless the traps are regularly inspected. If the trapper can work only at certain times of day, he had better put his traps out of action when he is not there, or use pull-the-string traps only. In any case a man-actuated type of trap should be the only one used in the breeding season, because no parent bird that is feeding young should ever be delayed for more than a minute. Live trapping, if it is properly and carefully done, does not do any harm at all to birds. Many trappers I know tell me of birds which get the trap habit, coming in sometimes three or four times in a day, perhaps under the impression that a free meal is worth a little inconvenience.

Handling Birds. – It is not very difficult to handle a bird (see Fig. 19), though some people seem to make a lot of fuss about it. You frighten the bird less if you are not frightened of it. If you are right-handed and want to put a ring on a bird's leg, you should grasp it with your left hand, your palm resting on its back. Its head should lie between your index and third fingers, which should curve round and rest gently, but not press, on its breast. Such few cases as I have heard of, of deaths of birds in the hand, have always been due to pressing of the breast with consequent crushing of the delicate organs in that region. When you have got your bird in the way I have described you can use your thumb and little finger to control its movements. In this you should have little trouble once you have turned the bird on its back; your thumb and little or fourth finger can then be used to hold one leg up for the ring. The instructions issued with the rings tell you whether or not you should overlap the size used for the species you are handling. Some small sizes of rings can be pressed on with the fingers, but the larger ones need a good pair of pliers.

When the ring is on, and its number noted, the bird should be quickly released. Very often your fingers can be opened and the bird will lie quiet on its back in your palm. As you tip your hand over it will recover from its hypnotic state and fly away.

Rings (see Fig. 20) should always be kept in order. They arrive in packets of twenty, and can be threaded on a thick string, or on a stick. You should keep a day notebook for rings only, and a card index under species and numbers of your own for your personal use.

British rings are issued and managed by the British Trust for Ornithology, to whom the scheme was lately entrusted by the proprietors of the magazine *British Birds*. At the present moment British rings can be purchased only by a member of the Trust or by a registered reader of *British Birds*. Details of the Trust will be found at the end of this book. *British Birds* is obtainable from Messrs H. F. & G. Witherby, 5, Warwick Court, London, W.C.1. Rings are issued by the scheme's organizer, Miss E. P. Leach, the Bird Room, British Museum (Natural History), Cromwell Road, S.W.7.

If you find a bird with a ring on its leg you should remove

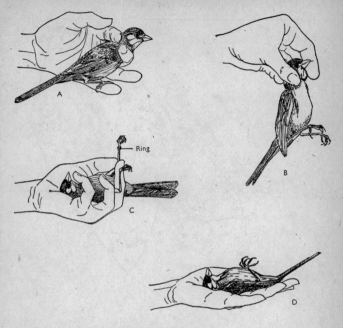

FIG. 19. How to handle a bird.

A. Perfectly safe handling; note second and third fingers round bird's neck.
B. It does not hurt a bird to be transferred from hand to hand thus.
C. A good position for ringing; note the gentle grasp of the ring-leg.
D. After handling, a bird will often stay still thus, not flying until tilted off the hand.

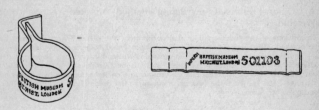

FIG. 20. A modern British Museum ring, designed for gannets and herons. The ends of some types simply meet or overlap, and have no clip turn-over. All bird rings are made of aluminium. The ring is shown in its normal shape, and spread out.

FIG. 21. Some results of the 'ringing' method. Adapted from Schüz and Weigold.

A. Migrations of the fieldfare. The marks represent the places where fieldfares have been marked with rings, or where they have been found with rings on their legs. The black squares refer to summer and the crosses to winter birds. It can be seen that the fieldfare is a summer visitor to northern and Alpine Europe and a winter visitor to Britain and south-west Europe. The continuous line encloses the area in which ringed birds have been known to be present in summer; the dotted line encloses the corresponding winter area.

the ring and send it to Miss Leach. If it bears the inscription 'Witherby High Holborn London' or 'British Museum London' it has almost certainly been placed on the bird in Britain.

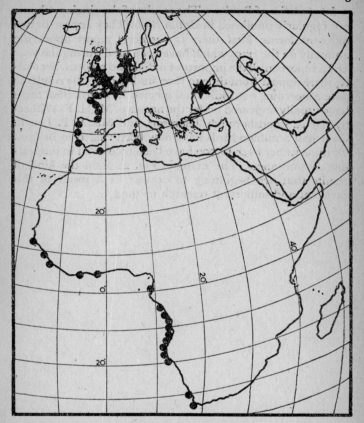

B. Migrations of the Sandwich tern. The stars represent the places where the terns
have been ringed in the breeding season and the dots the places of recovery. It can be
seen that the Sandwich tern is a trans-equatorial migrant, and that its chief route lies
down the west coast of Africa.

If it bears a foreign inscription, it should still be sent to Miss
Leach, because she handles all the foreign contacts and is
responsible for collecting all information about foreign birds
recovered in Britain.

Interpretation of Results. – The results of the ringing scheme are regularly published in *British Birds*. They appear in two different forms. First there is a general report appearing annually on the rings which have been used during the year and a report on all the important individual records. Secondly, there appear from time to time studies of all the results to date concerning one individual species. In *British Birds* these have usually appeared under the names of Mr H. F. Witherby, Dr A. Landsborough Thomson, and Miss E. P. Leach. Over half a million birds have now been ringed in Britain. Of these rather over 10,000 have been recovered in interesting circumstances. The maps in Fig. 21 show the kind of results that are beginning to appear as a result of this excellent and important research method.

CHAPTER V

Where Birds Live

❧

Selborne parish alone can and has exhibited at times more than half the birds that are ever seen in all Sweden; the former has produced more than one hundred and twenty species, the latter only two hundred and twenty-one. Let me add also that it has shown near half the species that were ever known in Great Britain.

GILBERT WHITE, 2 September 1774

❧

MOST creatures, except the very unspecialized and adaptable ones like man, are governed to a remarkable extent by the nature of their environment. Birds are no exception to this rule. A species of bird out of its geographical range or its habitat is as great an oddity as a quotation apart from its context. As this book is being written with special reference to the islands of Great Britain and Ireland, it might be as well to describe this part of the world from the bird's point of view.

The British Isles are very rich in bird life. From the point of view of animal geography, they lie at the edge of one of the most important of the great zoological regions – the Palaearctic Zone. This zone extends from Britain to Japan, and its animal (and bird) inhabitants are in the mass different from those in the other parts of the world. The great region of the Nearctic, which comprises the north of the American continent, has the closest affinities with the Palaearctic region, but in many respects the birds, particularly those of the land, are different.

Of the countries in the Palaearctic region Britain has probably a larger list of birds than any other of like area and climate. This is largely due to the facts that it lies athwart the west coast of Europe, directly in the path of a major migration route; that it is provided with many diverse habitats – large areas of undeveloped marshland being alone absent; that its northern oceanic cliff-bound coasts attract sea-birds and others with arctic affinities; and – by no means least – that bird-protection and conservation (with one or two unfortunate exceptions) are almost second nature to the modern British, and that only a small group of species are slaughtered for food.

From the point of view of climate, Britain is more favourable to birds than many other parts of the Palaearctic region. Though situated in a rather northerly latitude, it is kept warm in summer, and merely cool in winter, by the Gulf Stream; the atmosphere is not too dry, nor too moist; there are no violent extremes of drought or rainfall. As a result, it has a large population of resident birds, a large number of visitors which reach it in summer from tropical countries, and a large number which visit it in winter from the arctic (for even the arctic is cool in summer, and our winter climate is little different from this). The great number which visit our country on passage, in autumn and spring, do so because across Britain lies a reliable and obvious migration route.

Very little of the surface of Britain is as it was before civilization came. In the north conifers and birch have given way to moorland, in the south oak scrub to parkland, plantations, pasture, and crops. And it is the surface that matters to the birds. In historical times many must have come and many have gone, and many have changed their habits. The following table shows the main features of the plant-environment in which British birds now have to live.

A. *Woodland*:
  (1) Coniferous:
       (a) Pine (natural)
       (b) Plantations
  (2) Deciduous:
       (a) Birchwood
       (b) Oakwood – pure or mixed (sandy soil) and oak-birch heath

    (c) Beechwood – on loam (often with oak or ash) or on sand
    (d) Ashwood, often with oak, sometimes birch
    (e) Alder, sometimes with birch and willow species
    (f) Scrub, often hazel or ash
  (3) Mixed:
    (a) Plantations, various types
    (b) Beech on chalk with yew and ash
    (c) Birch with juniper.

Most of the above types can be planted or natural; with open or closed canopies; with or without secondary growth.

  B. *Park or Garden Land*:
    (4) Fringing woods and coppices
    (5) Parkland
    (6) Orchards
    (7) Gardens
  C. *Agricultural Land*:
    (8) Allotments, etc.
    (9) Arable land:
      (a) Roots
      (b) Cereals
    (10) Grassland
  D. *Heath and Moor*:
    (11) Lowland (rough pasture)
    (12) Upland (sometimes rough pasture)
  E. *Alpine*:
    (13) Mountain tops
  F. *Water Land*:
    (14) Flowing
      (a) Streams
      (b) Rivers
    (15) Placid:
      (a) Ponds and small lakes
      (b) Large lakes
    (16) Stagnant:
      (a) Mosses and bogs
      (b) Marshes and fens
  G. *Coastland*:
    (17) Salt Marsh
    (18) Dunes
    (19) Beaches
    (20) Cliffs
  H. *Man-land*:
    (21) Built-up areas

I must stress that this is a general classification of the habitats of Britain arranged as far as possible from what appears to be the bird's point of view. Of course there are many other kinds of habitat in the world with many other subdivisions. In the tropical forests alone it is possible to describe about eight layers, one on top of the other, with a special bird fauna to each.

But such extensions of habitats need not concern us here. Our task is to describe the places birds have to live in in temperate Britain and the kinds of birds which live in such places. The picture we get will at the same time give us a good working idea of the situation in North America, because in that land the places in nature are roughly the same, and though the birds that fill those places may be of different species, they do much the same biological work.

To take our first category, we find for instance that the coniferous forests of America, Europe, and Britain (in the latter the last relic of primeval pine forest in the Spey Valley) have birds breeding in them which are strikingly the same, as far as their adaptation to mode of life goes. A list might run:

American Forests	European Forests	British Forests
(1) Wood-grouse	Willow-grouse, Capercailzie	Capercailzie
(2) Grosbeaks	Grosbeaks	
(3) Siskins	Siskin (different species)	Siskin (as European)
(4) White-winged crossbill	Continental crossbill, Parrot-crossbill	Scottish crossbill
(5) Spotted wood-pecker	Spotted woodpecker, Green woodpecker	Spotted woodpecker
(6) Nutcracker	Nutcracker (different species)	
(7) Jays	Jay (different species)	Jay
(8) Chickadee	Continental crested tit	Scottish crested tit
(9) Kinglet	Goldcrest and firecrest	Goldcrest

The bird population of conifers in Britain is rather interesting, because except for those in the Spey Valley, practically all our coniferous woods belong to artificial plantations.

After the last Ice Age, perhaps 15,000 years ago, it is probable that pine and birch forest extended well down to the south of Britain. The birds that lived in these woods originally spread down with them; Lack and Venables[1] suggest that these were the crossbill, siskin, lesser redpoll, treecreeper, goldcrest, coal-tit, willow-tit, crested-tit, capercailzie, and black grouse. Later these woodlands moved north again (at the dawn of the Christian era it is probable that the only pine forests were within the Scottish Highlands), to be replaced by the typical oak forest of the south, and the typical conifer birds were left with two alternatives— to retreat with the conifers or to alter their habits and colonize the oak. Crossbills and crested tits and black grouse, among others, retreated, and the first two are now relics only in Scotland (for the crossbills that now breed in conifer *plantations*[2] in south Britain belong to the Continental race and have come over in irruption years). Tree-creepers, coal-tits, and gold-crests, however, have carved out a niche for themselves with some success in the oak woods; they also are typical birds, of course, of the southern conifer plantations which have been made by man in recent times. And into these planted pine-woods have also spread some birds of the broad-leaved woods; these plantations are seldom allowed to grow old, and in their early stages they are colonized by warblers, song-thrushes, robins, wrens, and hedge-sparrows.

Just as there is very little coniferous woodland left native to Britain, so there is very little deciduous. Practically all the woodland you see in the country to-day is planted. Yet whether it is planted or not, we can recognize many types,

1. See David Lack and L. S. V. Venables, 'The Habitat Distribution of British Woodland Birds', in *The Journal of Animal Ecology*, vol. 8, pp. 39–71 (1939).

2. Real plantations of pine and other conifers (as distinct from ornamental groups) were not made in this country until the beginning of the eighteenth century; it is interesting to note that the irruptions of crossbills from the Continent noted by Gilbert White in that century were followed, in the next, by the beginning of a slow colonization—see Fig. 16.

each having a fairly distinct bird fauna. An important group of deciduous woodland is what can be generally called Northern Scrub. It is often composed of alder, often with birch and many of the different species of willow, and in the untouched northern regions it forms a fringe between the edge of the coniferous forests and the real tundra. Iceland and Scandinavia have much of this scrub, but in Britain it has nearly all been cut or grazed away. Among the typical birds of this northern scrub-land are finches like redpolls and thrushes like redwings. There are still plenty of redpolls nesting in Britain, but with the loss and cutting away of most of their proper scrub-land, the redwings have gone (probably some hundreds of years ago) and to-day the discovery of a redwing nesting in Scotland is usually kept confidential.

The main advantage which deciduous woodland has over coniferous is, from the bird's point of view, the presence of numerous holes in which to nest. The anatomy of a broad-leafed tree tends to permit the development of holes, whereas that of a narrow-leafed tree almost prohibits this. This is perhaps one of the main reasons why the bird population of deciduous woodland is usually more than twice that of pure conifer woodland.

Mixed woods, plantations of perhaps beech and pine, spruce and oak, can be very attractive to bird life, because of the very large surface area such plantations usually have. A wood commends itself to birds, not by having particular types of trees so much as by having a particular structure. The important factors are height of main growth, height of secondary growth, whether the canopy is open or closed, whether there are rides and other spaces through the woods, and so on. An open canopy is one which, when the trees are in full leaf in summer, admits light in large quantities through to the ground and the secondary growth. A closed canopy is one in which the light is shut off from the ground. Generally speaking, the more open a wood and the more exposed the secondary growth, the higher the bird population. Many sorts of warblers will sing from song posts on the high trees and dodge down through the open canopy to visit their nests in the secondary growth.

Perhaps the highest bird populations of all are found in areas where man has interfered with and controlled the distribution and numbers of trees (see section B of the table on p. 87). The numbers of breeding birds are always high in an area where there is great variety of cover and choice, and in a garden in which there are tall trees, shrubs, bushes, undergrowth, nest boxes, eaves, water, and so on, there are likely to be many nest sites, so that birds may breed in the garden but forage outside. This island effect has produced some very high populations in gardens.

Fringing woods, parks, coppices, orchards, and gardens are artificial and man-made things, and by no means constitute the environment to which most of our British birds are naturally adapted. We find these habitats have been occupied in Britain by such birds as are fairly plastic in their selection of where to live. Blackbirds and robins were probably originally birds of the primeval deciduous woodland, which were not so specialized physically or psychologically that they could not adapt themselves to the new circumstances. Bullfinches, hawfinches, and green-finches came to the park and garden land from scrub, pied wagtails probably from water meadows.

Before we leave the subject of trees, it is important to note that they form important breeding sites for birds which otherwise live in open country. This is perhaps not quite so noticeable in Britain as in parts of the great central Continental grasslands, where islands and clumps of trees are full of birds which breed there, but which spread right out through the plainland when they are foraging. You cannot have rooks feeding in your open pastures and arable land unless there are woods for them to roost in in winter and spinneys and coppices for their summer rookeries. The only resident rook in Shetland lately was a bird that was blown there and was tamed by a crofter. There is plenty for this rook to eat, but nowhere for it to breed. The most northerly rookery in Britain is in the most northerly wood, a clump of trees in a sheltered place in Orkney.

On agricultural land birds tend to feed rather than to breed. But in Britain such land is bounded and divided by hedges which are so abundant (perhaps a mile of hedge to every forty acres) as to have a very marked effect on the bird population. Every spring the breeding-places provided by these hedges are exploited to the full by buntings, finches, hedge-sparrows, robins, wrens, blackbirds, and thrushes. A few birds like larks find nest sites on the ground in the meadows, but it is the hedges which keep the population as high as that of coniferous woodland and almost three times as high as that of moorland, heath, and rough pasture. On a typical Oxfordshire farm the species with a summer population of more than three birds to 100 acres were starling, blackbird, chaffinch, house-sparrow, linnet, skylark, yellow-hammer, songthrush, robin, rook, lapwing, wood-pigeon, hedge-sparrow, wren, and whitethroat.

There are few birds on heaths and moorland, though one or two of them are specialized and well adapted to their place in nature. Only the skylark, meadow-pipit, red grouse, and, in parts of Scotland, the oyster-catcher can muster an average population of over three birds to 100 acres – compare this with farm-land. We have to go down to a density of three birds to 1,000 acres to bring in such birds as the twite, linnet, reed-bunting, tree-pipit, wheatear, hedge-sparrow, ringed plover, golden plover, lapwing, curlew, and partridge.

Some of the birds which breed on the moors, like curlews, dunlin, golden plover, have, as their winter places, the wild shores and mudflats of the coast. Some of these are only just hanging on to their breeding-grounds; to-day dotterels nest only on a few remote uplands, and whimbrel only in certain rather secret northern isles.

All these moorland birds have to nest on the ground or in such low bushes and whins as they may find. Even moorland birds of prey like merlins have to nest on the ground.

On the very high mountain tops moorland and heath give

way to an Alpine vegetation with low creeping willows and saxifrages and much moss. Here the corner or crevice in the rock becomes the nesting site, and its inhabitants the ptarmigan or snow-bunting; sometimes the twite. These are probably birds which have had to retreat up the mountains since the world got warm after the last ice age. The ptarmigan and snow-bunting have been almost driven out of Britain altogether, for they only survive on the tops of the higher Scottish peaks.

You will not find dippers or grey wagtails away from streams. The dipper is adapted in a quite remarkable way for a life in swift water. The tilt of its back against the running stream enables its feet to get a purchase on the bottom and control the direction of its movements in its search for the larvae of water creatures. Both the dipper and the grey wagtail nest in crevices and holes, in banks or under bridges, never far from their water.

One could prepare innumerable lists of birds and the habitats to which they are restricted. One can assign kingfishers and grey wagtails to rivers and streams, coots and red-throated divers to ponds and small lakes, many species of duck to larger lakes, herons and bitterns and bearded tits to marshes and fens. On the coast a salt marsh, sand dune, beach, and cliff have each their own distinctive fauna. The house-sparrow is as much a symbiote of man as the jackal is of the tiger.

But before we leave the subject of the habitat, we must suggest that its study in detail means much attention to the psychology and behaviour of birds. The very interesting work of Lack and others on the reasons why birds are disposed to select one habitat rather than another has already made it clear that idiosyncrasies of mind as well as of structure play a great part in determining where a bird is to live.

Perhaps the most interesting thing that becomes clear is that within one gross habitat like woodland there is a sort of share-out between the birds. Psychological preferences for things like the height of song posts, open or closed woodland, type of nest, site, and material and so on may be as important in determining a bird's choice of world as his food preferences or tolerance of climate. As our study of these problems increases, we will I think be sure to find that these psychological barriers and preferences, which birds have set up, mean that from their standpoint the gross habitats can be divided up in such a way that more species could be supported by the available food than would otherwise be the case.

CHAPTER VI

The Numbers of Birds

❦

Among the many singularities attending those amusing birds, the swifts, I am now confirmed in the opinion that we have every year the same number of pairs invariably; at least the result of my inquiry has been exactly the same for a long time past. The swallows and martins are so numerous, and so widely distributed over the village, that it is hardly possible to recount them; while the swifts, though they do not build in the church, yet so frequently haunt it, and play and rendezvous round it, that they are easily enumerated.

GILBERT WHITE, 13 May 1778

❦

THERE are about a hundred thousand million birds in the world; that is, there are probably a hundred thousand million rather than a million million or ten thousand million. Two attempts have been made to estimate the total May land-bird breeding population of Britain. The first is that of E. M. Nicholson, who, in 1932,[1] put the figure for England and Wales at about 80 millions. My own preliminary figure for Great Britain, excluding Ireland, was about 100 millions (1939).[2] The table shows a detailed revised assessment of these figures made from the latest information available. This puts the population at about 120 millions.

Since there are large areas of moorland in Scotland and the bird population is consequently low, it can be seen that Nicholson's figures agree closely with mine; certainly he must have the credit for being the first to give an estimate for the number of land birds in Britain. Both our estimates

1. *Daily Mail*, 29 September.
2. *Birds as Animals*, 1939, London (Heinemann).

are based, as can be seen from the table, on sample counts of breeding birds taken in the more important habitats. Much of the census work has been done by the British Trust for Ornithology.

Habitat	No. of thous. acres 1939	Birds per 10 acres (approx.)	Total No. of birds (approx.)
Cereals	6,391		12,782,000
Roots, 'cabbage', lucerne, etc.	2,161		4,322,000
Clover, etc., and planted grasses	3,361	20	6,722,000
Other crops	267		534,000
Bare fallow	370		740,000
Permanent grass	17,411		34,822,000
Rough grazing	16,081		11,256,700
Deer forest	3,430	7	2,401,000
Other ungrazed moorland	230		161,000
Parks, golf-links, etc.	33	100	330,000
Orchards	253	300	7,590,000
Small fruit	58	300	1,740,000
Hops	19	100	190,000
Gardens, allotments, etc.	313	300	9,390,000
Woodland:			
Coniferous	672	20	1,344,000
Deciduous	443	40	1,772,000
Mixed	301		1,505,000
Fringing and coppice	734		3,670,000
Waste, scrubland, and swamp	939	50	4,695,000
Built-up*	2,740		13,700,000
Inland water	593	10	593,000
TOTAL	56,800		120,259,700

* Approximate.

Though there are 426 species in the British list, only about 100 of them contribute anything significant to this total and the great bulk of it (about 75 per cent) is made up by thirty species, all of which have a population of 350,000 or over in England and Wales. The commonest birds in England and Wales are the chaffinch and blackbird, of which there are about 10 millions each. There are about 7 million starlings, and the same number of robins. House-sparrows, hedge-sparrows, song-thrushes, and meadow-pipits number about 3 millions each, rooks 1¾ millions. At about 1¼ millions we find yellowhammers, wrens, whitethroats, willow-warblers, and wood-pigeons; at three-quarters of a million, jackdaws, skylarks, blue tits, swallows, house-martins, and linnets; at 350,000, green-finches, great-tits, tree-pipits, mallards, chiffchaffs, moorhens, swifts, lapwings, and partridges. It should be stressed that these are very approximate figures.

These are the birds with large populations – so large that they can be arrived at with some degree of accuracy from taking counts of random samples of the countryside. With birds of smaller populations the numbers are so few that, unless one takes a very large sample, the total population cannot be accurately estimated. Some other method must be invented for counting them. As far as land birds go, it has been possible to count only a few with small populations. In England and Wales there are about 150,000 black-headed gulls (these are included among land birds, as many of them breed miles inland), approximately 25,000 barn-owls, about 8,000 herons, and about 2,650 adult great crested grebes. The heron population of *c.* 8,000 birds is that of good years; severe winters such as that of 1947 may cut it in half.

It has been possible to count these birds because of the existence of co-operative observers in all parts of the country (chiefly members of the British Trust) and because of the particular habits of the birds themselves. Herons and black-headed gulls nest in very obvious and sometimes spectacular colonies, those of the latter in one case reaching nearly 50,000 pairs (Ravenglass, in Cumberland, which has about two-thirds of England and Wales' population). In many areas where there are more than a certain number of observers, these colonies can all be found and counted, and this has been done in these particular cases, though the black-headed gull figures have had to receive certain mathematical treatment before they could be finally estimated.

The only other land birds whose population can be arrived at with any accuracy have up to now been those which are extremely rare. Quite often a pair or two of birds are the only ones known to be breeding in the country and one can arrive at such obvious figures as two for the population of the black-necked grebe in England in 1939, about twenty for its population in Scotland, and under 250 for its population in Ireland. At present about sixty black redstarts breed in England, and nearly a hundred little ringed plovers. Kites are to-day breeding only in Wales, where the latest survey gives a figure of about twenty-four birds and not more than six nests. Probably not more than four marsh-harriers and fifty Montagu's harriers are breeding in England, and not more than about fifty hen-harriers in Scotland. I should not be surprised if there were no more than 250 pairs of golden eagles in Scotland. In the half-century 1900–49 in England and Wales, the golden oriole, hoopoe, and honey-buzzard have probably nested less than ten times, the tawny pipit twice, the moustached warbler once, the icterine warbler and the little bittern once or twice, the red-crested pochard once, the goldeneye twice, the black-tailed godwit and green sandpiper once, the ruff at least six times, the black-winged stilt thrice (in one year). About fifty avocets or more now (1950) breed in England, though the species was extinct for over a hundred years and did not recolonize the country until 1946 (two pairs bred in Ireland in 1938).

One rare subspecies has a population which is pretty accurately known. This is the St Kilda wren. In 1931 the Oxford Expedition recorded a total of sixty-eight pairs of this interesting little bird, and my subsequent visits in 1939, 1947, 1948, and 1949 prove that no great change has taken place in its numbers.

The numbers of sea-birds are extremely large, probably as large in Britain as those of birds inland. Nobody knows much about the actual numbers of these birds except of certain rare species and others which have been the subject of specially intensive study. A rare bird, Leach's fork-tailed petrel, breeds in only three or four islands off our coast and probably cannot muster more than 4,000 individuals, though the true population is difficult to assess except at night, when the number can be roughly estimated by the sounds made by the incubating birds. In England and Wales, where the greater black-backed gull is fairly rare as a breeding species, there are about 2,000 breeding adults, though the population in Scotland and Ireland is large relative to this, and no attempt has so far been made to assess it.

By means of a partly indirect mathematical calculation I have been able to put the population (in 1949) of breeding fulmars in the whole of Great Britain and Ireland at rather over two hundred thousand – about five times what it was fifty years ago. There is only one bird whose world numbers are not small and which are anything like accurately known. In 1939 a group of ornithologists, including H. G. Vevers and myself, visited nineteen out of the twenty-two gannet colonies in the world, and were able to make counts of all but 2 per cent of the birds.[1] We believe that there were about 165,600 gannets breeding in the world in 1939, of which 109,100 bred in Great Britain and Ireland, and about 11,800 in England and Wales. All but about eight of the latter bred on the colony of Grassholm, in Wales. The eight came from what appears to be a new colony that is trying to establish itself on the Yorkshire coast. Fig. 22 (p. 104) shows the 1939 world distribution of the gannet, and gives figures for the numbers of pairs of birds at each colony. In 1949 we repeated the census of most of the colonies and found the population to have risen to about two hundred thousand.

Besides the gannet there appear to be only eight other

1. Since reduced to $1\frac{1}{2}$ per cent by a visit in 1940 to one of the three colonies not seen in 1939.

species of birds whose world population is known. These are rare and possibly on the verge of extinction. To-day there are probably no more than 600 great white herons or 500 trumpeter swans. There are about 55 Chatham Island robins, 50 of the recently rediscovered takahe of New Zealand, thought to be extinct, 45 California condors, 40 whooping cranes, 24 Hawaiian geese, and perhaps a dozen ivory-billed woodpeckers. I do not see much chance of our being able to establish the world populations of any others at all accurately, except under one or more of the following conditions.

(a) That they are social sea birds with a limited number of colonies and a definite breeding season, so that expeditions can count the number of nests at each colony.

(b) That they are social birds with a very restricted geographical distribution, and a limited number of colonies irrespective of whether they are sea birds.

(c) That they are large, spectacular, and easily recognizable birds (like the trumpeter swan and California condor) which are nearly extinct.

(d) That they are species of birds restricted to small islands or isolated areas. Thus there are probably between one and two thousand flightless rails Atlantisia on Inaccessible island in the South Atlantic; not more than two thousand Kirtland's warblers in the restricted jack-pine area of Michigan; not more than two thousand Ross's geese nesting by the lakes of the Perry River in Northern Canada.

Here it must be mentioned that the endemic species of the Galapagos Islands, off the west coast of Ecuador, have small populations. Such birds are the Galapagos penguin, flightless cormorant, albatross, flamingo, and perhaps one or two of the many local species of Darwin's black ground finches. It might well be possible to count some of these.

The Commonest Bird in the World. – Darwin once suggested that the fulmar was the commonest bird in the world, but judging from its world breeding-distribution, I should be inclined to disagree with him. Modern research has shown that this distribution is far more restricted than was previously supposed. Certainly the number of fulmars must not be judged from the outpost population in the British Isles, large though it has now become. In Iceland and the Faeroes, Jan Mayen, Bear Island, Spitsbergen, and Greenland there are many thousands of fulmars. There may even be several millions, but I should very much doubt whether there are ten millions. Darwin may have got his idea from the abundance of the fulmar at sea in the North Atlantic. Certainly it is one of the commonest birds in some parts of that ocean, but I would suggest that the little auk, another arctic breeding species, runs the fulmar fairly close if it does not actually beat it. Little auks breed in fewer places than fulmars, but in enormous, closely packed colonies on such arctic lands as they inhabit.

It is very difficult to assess large numbers. It needs a great mental effort to think in the necessary terms, and comparative numbers become as hard to imagine as actual ones. All the same, the most abundant bird in the world is certainly a sea bird, and probably Wilson's petrel. Curiously enough this creature has been observed in Britain only about a dozen times. Its breeding haunts are the Antarctic and some of the surrounding islands, and here its numbers, Dr B. B. Roberts tells me, are quite fantastic. In the off season, Wilson's petrel moves into the oceans, and huge flocks pass up the Atlantic, Pacific, and Indian Oceans, and even into the Red Sea. As they get north they tend to congregate on the west side, reaching their peak in numbers in our northern autumn in the neighbourhood of Long Island Sound. There are probably one or two other sea birds which may run Wilson's petrel fairly close. These are the cormorants of the Cape and of the Guano Islands of Peru.

It is possible to guess at the most abundant land bird. Without doubt it lies between the starling and the housesparrow. Both these birds have origins in the European continent, where they are very widespread, and have been transferred by human agency to other continents. In North

America both starlings and house-sparrows are pushing rapidly westwards, where they are in many regions dominant birds. The house-sparrow is getting to other parts of the world as well. It has reached Uruguay and arrived at the Falkland Islands in 1919. It was introduced into New Zealand in 1862, and into South Africa at the end of the nineteenth century. Though the house-sparrow has now a wider distribution in the world, the starling has probably the higher population, since it is not so restricted in its habitat. It must be stressed that in Britain, where the starling ranges over wide areas of countryside, it is over twice as numerous as the house-sparrow. This fact might disturb some people until they realize that they are likely to know more about the neighbourhood of houses (and therefore come across the house-sparrow) than about the general nature of the open country.

The most widely spread land bird in the world is pretty certainly the barn-owl. Its races extend to every part of the world except the polar regions and New Zealand and certain of the Pacific Islands; through these last it has penetrated as far as Fiji.

Methods of Counting. – There are two main methods of counting birds, the area census and the breeding count of an individual species. Both these methods are valuable for finding actual populations and for comparing the situation in different years. I must again emphasize that our general idea of land-bird breeding population has been arrived at from the first method, which is perhaps the more important. The general technique is to take a sample of one of the major habitats, to find the area of this sample, and to find the total number of breeding pairs of all species in it. In the long run the only really accurate method is to find every nest, a method which needs much skill and practice.

But rough figures can be arrived at without this prodigious exercise. If the visit is made at the right time of year, the presence of a breeding pair is often indicated by territorial song. And once the numbers of one or two species have been found by this method, the numbers of the others can be roughly assessed from the relative numbers of times the different kinds of birds are seen or heard.

Most of the intensive work on the bird life of woodland, moorland, and so on in Britain, done by members of the British Trust, has been of this comparative kind, and we have a picture of the relative abundance of the different kinds of birds which is a good deal more accurate than our idea of their actual numbers. Of course, relative numbers are of great biological importance, so it is not surprising that much work and thought has been given to their discovery. The simplest method is to walk through a selected part of the habitat and make a tally-list of the species and the numbers of each seen or heard.

Another method is to visit the habitat on several successive days and to note, not so much the numbers of each species met with on one visit, as whether or not the species has been met with at all on that day. When the visits have finished, the total number of days on which each species has been seen gives a very good index of its relative abundance.

It can be seen that in area census work there is a lot of scope for individual as well as collective work. A sample count can be taken by the solitary observer walking quietly through a piece of woodland, for instance, and would probably be more accurate than if he was accompanied by a talkative companion. On the other hand, the actual rather than relative numbers of birds in the same wood might well be discovered by a team of bird-watchers of roughly equal calibre walking in line, abreast, at twenty- or thirty-yard intervals. On low grassland these observers might usefully trail a rope between them so as to disturb nesting birds and to ensure that the scheduled acreage is thoroughly worked. Nevertheless, even when the most thorough search of the area is made, some species manage to elude the observer, while others remain peculiarly conspicuous; and much re-

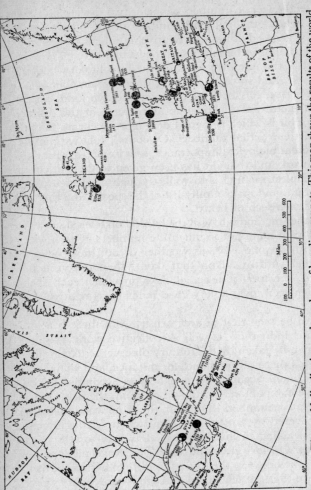

FIG. 22. The world distribution and numbers of breeding gannets. This map shows the results of the world gannet census in 1939. All the known colonies are marked and the number of occupied nests (which is approximately the same as the number of *pairs*) at each colony is shown. Circles with dotted centres show the sites of extinct colonies; at Great Saltee about six birds were about but no nest was found.

search remains to be done before a method of assessing the true population of land birds is perfected.

Counting the Individual Birds. – As I have already suggested, most of the counts of individual species of birds have been successful when the birds have been social breeders. This particularly applies to sea birds. The main task of the gannet-counting teams during their surveys in 1939 and 1949 was not so much the actual counting as getting to the colonies. In this particular case the colonies are on the remotest possible stacks and rocks (see Fig. 22) like Eldey, 20 miles from the nearest fishing village out in the Atlantic off the coast of Iceland. St Kilda is 45 miles west of the Outer Hebrides. Sula Sgeir is about the same distance north-west of Cape Wrath. Sule Stack is about 35 miles from Cape Wrath and the same distance from Orkney. The Hermaness (Muckle Flugga) colony is the northernmost tip of the British Isles. Once the colonies have been reached, there are certain further technical difficulties, such as those of landing and climbing, but the task of counting each occupied nest is comparatively simple, though it imposes a certain strain on the eyes.

Most people who visit a place where birds are breeding carry away some impression of population. Usually this is interpreted in their notebooks as few, common, abundant, and so on. It needs little extra effort to make these note entries less vague, and one's rapid impression can quite easily be translated into terms of under 10, under 100, under 1,000 (and so on) breeding pairs or breeding birds. Of course, notes on actual numbers are far more useful than notes of the above kind, but those general figures are far better than no figures at all, and can often be interpreted to a good deal of advantage by somebody who is working up the particular species later. Our knowledge of the fulmar's increase in population (as opposed to its simple geographical spread) is largely due to interpretation of observations in terms of under ten and so on.

The numbers of birds fluctuate, often in an orderly and

remarkable way. In Chapter IV we have discussed some of these fluctuations, especially those of species like the cross-bill and Pallas's sand-grouse.[1]

But there are many trends in the numbers of British birds which we cannot yet ascribe to periodic fluctuation. Some species are tending to increase and spread, others to decrease and disappear from their previous haunts.

Spreads. – Perhaps the most spectacular spread in recent years has been that of the fulmar. In the year 1697 this bird was known to breed on the island of St Kilda, and though there are one or two early records of colonies elsewhere, none of these can be really substantiated. In 1878, however, a colony was established in Shetland on the island of Foula, and since that date the number of known colonies has in-creased to such an extent that in 1939, 208 different stations were known at which birds were definitely breeding, and 61 others at which birds were present in the breeding season but had not yet been proved to breed. By 1949 there were 365 known breeding colonies and no less than 208 prospect-ing stations. The spread took place first to the more oceanic headlands, followed by the colonization of the intermediate cliffs, and a gradual pushing up the firths and lochs. Down the coast of Britain fulmars reached the limit of high cliffs in Yorkshire in 1922, by 1940 were prospecting the lower cliffs of Norfolk, and in the breeding season of 1944 were seen at cliffs in Kent. In 1944, too, they first bred in Cornwall and on Lundy Island off the Devon coast. They have been recently seen, in summer, at cliffs in Dorset and Sussex and now breed in South Devon. They now breed in Pembroke-shire; have pushed down the North Channel into the Irish Sea, and breed on the Isle of Man and in Dublin county. They bred in 1945 on Great Orme's Head. In Cumberland

1. A great deal about fluctuation in the numbers of animals can be learnt from Charles Elton's *Animal Ecology* (London, 1927), and his paper in the *British Journal of Experimental Biology*, vol. 2, pp. 119–63, on 'Periodic Fluctuations in the Numbers of Animals: Their Causes and Effects', is well worth the trouble of finding.

they first bred in 1941, at St Bee's Head. They have pushed up the Inner Hebrides from the south. The series of maps in Fig. 23 shows roughly the way their spread has taken place.

Most of the colonies, except for some in the very north of Britain, are still small; their numbers have increased much more rapidly than the total population of the bird, which has probably quintupled in its seventy years of imperialism. This multiplying is, all the same, remarkable, since the fulmar usually lays but one egg a year and it never replaces it if it is broken or lost, and it probably takes eight years or more to reach maturity. Up to 1914 the increase was at about 18 per cent compound interest, and after that time has been at over 7 per cent per annum.

It was recently suspected that the increase might have been due to an overflow from St Kilda, since at roughly the time it started the St Kildans began to receive supplies of food from the mainland and were supposed to have discontinued their famous wild-fowling habits. But an exhaustive survey of the literature about St Kilda, which is very complete, shows that until 1910 at least the St Kildans went on taking birds as much as ever. Moreover, it is becoming clear that the fulmar started spreading many years before the first new colony was established in Britain. The spread dates as a matter of fact from about 1753, when the bird began to increase greatly in Iceland, and some time between 1816 and 1839 it arrived at the Faeroes, north-west of Shetland. In the Faeroes it spread so much that recently the inhabitants were taking and eating 100,000 young a year, which is very much more than the total number of young produced in one year in Britain. The spread is probably due to the northern whaling industry, and after it the trawling industry, which have altered the fulmar's food-supply by providing large amounts of animal waste.

There have been many other recent and remarkable spreads of species of birds in Britain. The little owl, which was introduced into Britain (in Northamptonshire) in 1889

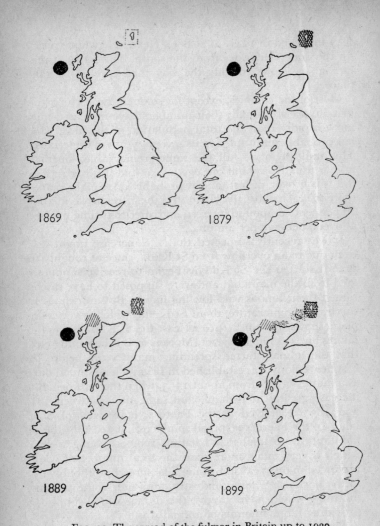

FIG. 23. The spread of the fulmar in Britain up to 1939.

A. Series of maps showing the broad distribution and population every tenth year since 1869. (For vice-counties see Fig. 14, p 57.)

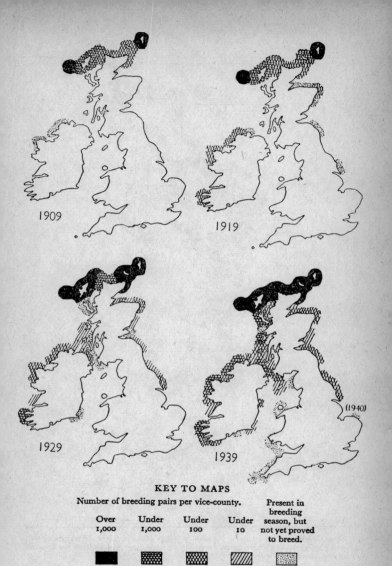

1909

1919

1929

1939

(1940)

KEY TO MAPS

Number of breeding pairs per vice-county.

| Over 1,000 | Under 1,000 | Under 100 | Under 10 | Present in breeding season, but not yet proved to breed. |

B. Map showing all the known colonies in 1939. Black circles: breeding-colonies; crossed circles: stations where fulmars were then prospecting. Arrows show the approximate directions of the spread.

THE NUMBERS OF BIRDS

and in Kent in about 1874 and again in 1896, has now spread to every corner of our country, as far north as Lancashire and Yorkshire. It reached Norfolk in 1912, Land's End in 1923, Pembrokeshire by 1920, and the Pennines and North Wales by 1930. Its natural distribution was west in Europe to the Channel; but owing to its resident habits it had not (save for a few vagrants which did not establish themselves) 'discovered' England. Once a breeding stock had been imported, however, it found that the place it occupied in nature in its native haunts was not occupied strongly by any competitor.

There have been some signs of recent invasion from the Continent by one or two species which seem to be generally extending their range in the west. The black redstart seems to be arriving in increasing numbers year by year, and nearly every season brings new records of its breeding in southern counties. It has bred in the heart of London since 1940, and has lately prospected Lancashire and Yorkshire. The stockdove has invaded Ireland presumably from England and Wales: since 1875, when it was first noted, it has increased and spread, till to-day it is breeding in nearly every Irish county. At the same time this bird pushed into the north of Britain and invaded Scotland. To-day it breeds in the west as far as Argyll and on the east right up to Sutherland.

The great spotted woodpecker is another bird which is pushing its range forward. In recent years it has become very much more common in the north of Scotland. In 1939 it was definitely recorded from a wood in Sutherland not far from the Caithness border, and lately it has been spreading through Argyll and into Western Ross (see Fig. 24).

I could quote many other examples of increase of range; it is particularly noticeable among many woodland and garden birds, many ducks and many sea-birds. Some species are finding new habitats as well as new lands to conquer. Thus curlews have increased and spread on to relatively low ground in England in the last thirty years. Black-headed gulls are tending to breed more widely inland; and are to-day land birds as much as sea birds. Their numbers are increasing too (see Fig. 25). In the north of England and Scot-

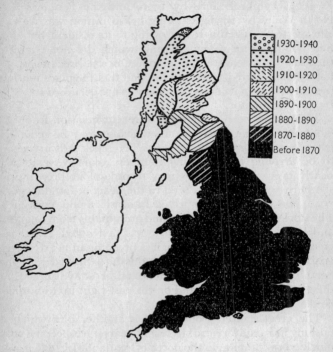

∘∘∘	1930-1940
⋯	1920-1930
⟍⟍	1910-1920
⟋⟋	1900-1910
☰	1890-1900
	1880-1890
	1870-1880
	Before 1870

FIG. 24. The spread of the great spotted woodpecker in the north of England and Scotland. Our knowledge of the spread dates from about 1870, and the shaded parts of the map indicate the bird's advance in each ten-year interval since then. From information compiled by H. F. Witherby (see *The Handbook of British Birds*).

land, oyster-catchers are spreading up the river valleys and breeding many miles away from the sea-shore. The green woodpecker is showing an increasing tendency to desert woods and forage in open fields.

Decreases. – Some of the causes of spread are hard to define. So are some of the causes of decrease. We just do not know why wrynecks have become so increasingly rare in the last few years, or why rock-doves are now no longer breeding in England or Wales, or on the east coast of Scotland south of the Firth of Forth.

Many of the decreases of British birds are, however, attributable to human agency. These causes, it should be stated,

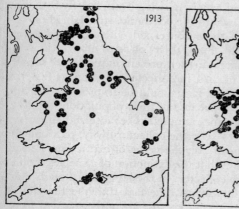

FIG. 25. Black-headed gull colonies in England and Wales in 1913 and 1938, showing the increase in number and the spread inland. Adapted from maps by P. A. D. Hollom.

are in the long run often deep-seated and sometimes out of the control of the protectionists and conservationists. It was the draining of the fens and not the absence of bird protection laws which lost us avocets, cranes, spoonbills, ruffs, and black-tailed godwits as breeding species. The cutting up and working of plains-land by man drove out the great bustard as a breeding species by 1833, and is now making the stone-curlew a rarish bird, breeding on such barren downs as it can still find. It is highly likely that one of the most spectacular recent decreases, that of the corncrake, or land-rail, is due largely to human agency; the results of the British Trust for Ornithology's investigation organized by C. A. Norris (in which over 3,000 observers took part) show that the development of the modern mowing-machine, together with the somewhat earlier date of hay-cutting, has been very closely linked with the corncrake's disappearance. A large number of corncrake's eggs and young are destroyed by the mowing, except in the remoter parts of the Highlands of Scotland and Ireland, where mowing is still largely by hand, and also later, and where the corncrake still flourishes.

The main task of those concerned with bird protection seems to be threefold. First, their job is to ensure the continuance of certain areas in their present natural state so as to preserve a sample of bird life affected by man as little as possible. Secondly, their task should be to preserve the different species of birds because of their population, interest, and biological importance as much as because of their rarity. Thirdly, they have to keep the collector away, particularly from such species as are on the verge of extinction. I need fear nobody in saying that the extinction of a species should never be ascribed to collectors in the long run, though they may often be the last straw. The fact that the remaining half-dozen nests of kites to-day are jealously guarded from collectors by a private army of enthusiasts certainly does not mean that collectors alone have brought matters to their present state. Collectors probably got the last osprey's egg ever laid in this country, but cannot be altogether blamed for the fact that it was the last. The trouble is that when for various reasons a bird tends to become extremely rare, its importance and the importance of its eggs from the collector's point of view become correspondingly greater, so that its end is hastened. If collectors were finally prevented from breaking the law, as I hope they will be, it is possible that a remnant of many species now nearly extinct would be saved. But it is equally probable that this remnant would seldom become anything more than a remnant, and that while it would remain very interesting from a humanitarian aspect, it would never play a serious biological part in our bird fauna.

I do not think that the kite in Wales is certain to improve, even if collectors are kept away from the few nests that are left; the birds have survived at roughly their present population for fifty years, and may not suffer from in-breeding; but the young produced hardly ever seem to get a chance to establish themselves in new areas, perhaps because they are easy to shoot and trap. Many landlords and their keepers have still to learn to ask questions first, and shoot afterwards.

Bird Protection. – In this country we have inherited a wonderful bird fauna. It is rich in species, rare and common, of simple and curious habits, harmful and beneficial, dull and exciting. When we consider the preservation and conservation of this heritage of birds, we must have several attitudes in mind, and we must be careful that our special fancy for any one of these attitudes does not lead us to neglect the weight of the others. Our attitudes must be biological, scientific, economic, and aesthetic. We must allow ourselves to be sentimental only in so much as it allows us to satisfy the other more important attitudes. We must see ourselves first as animals living in this country in an ecological relationship with the wild creatures around us.

If a 500-years-old tradition encourages men from the Isle of Lewis to make a hazardous boat journey every year to take young gannets on Sula Sgeir, we must try to arrange that they are able to pursue this important and economic custom. We must not say this killing shall stop; neither must we say that we can leave the future of these gannets to chance. Our solution is to apply what knowledge we have of the vital statistics of the gannet, and to ration the numbers taken for the benefit of both man and bird. The men of the Faeroes, who take gannets every year from the Holm of Myggenaes, farm their birds as carefully as if they were chickens.

Most of the common small birds of the countryside are neutral from the economic point of view; there is precious little evidence that the insect-eaters do any good by insect-eating, or that the seed-eaters do any harm. The provedly harmful birds are few.

During the Second World War, when more corn was grown in Britain than ever, I organized an investigation of the numbers and food of the rook; it ate more corn than any other food, on an average, through the year. Nevertheless, the total amount taken was small, as far as our economy was concerned, and half of it was waste from stubble, which was of practically no interest to the farmer. The rooks did not eat as many insects as I had expected, and ate practically no wireworms. Even if they had eaten twice as many insects as they did, there is no evidence that this would have been of value.

The most injurious bird is certainly the wood-pigeon. Its depredations among corn-crops and greens are serious most of the year round.

It can be seen that there is no strong argument for protecting birds on economic grounds. In Britain, however, the most important reasons for protection have always been aesthetic. Among the many facets of the aesthetic argument, a new one is becoming important – that birds should be preserved, not only because of their rarity, curiosity, wildness, or romance, but also because of their biological interest. This new argument contradicts some of the old ones, for many of the predatory birds disliked by sportsmen and sentimentalists are of primary interest to biologists. The new biological conservation desires very little more than the effective removal of man, alone among the predators, from a living association of birds.

There are many signs in Britain of a reorientation of the efforts of bird protectionists, and societies such as the Royal Society for the Protection of Birds. Organized bird protection has now existed in Britain for over sixty years; long enough to have had a measurable effect on the British avifauna. There is a strong case for concluding that the recovery of our birds during the first half of the twentieth century, with the noticeable increase in both numbers and variety, is due not entirely to the amelioration of the climate but largely to better protection – to the protection of habitats and maintenance of sanctuaries, and to the improvement in public opinion. But sanctuaries for birds are not created simply by buying a place and calling it a sanctuary, nor is the public any longer persuaded by sentimental exaggerations of the economic value of birds or of the wickedness of collectors (though some collectors are as ruthless as they are stupid, and still wantonly endanger some very rare birds).

It is interesting to see that the old-established Royal Society for the Protection of Birds is steadily improving its sanctuaries (it has recently purchased Grassholm, one of the largest gannetries in the world, and an island on the Suffolk coast which is the chief headquarters of the avocets that have now so happily recolonized Britain) and its propaganda (*Bird Notes* is now published in a new format and contains many interesting and important articles). The most important step forward that can be immediately taken, in my opinion, is in the improvement of the law. The law at present is very involved. At least nine Wild Birds' Protection Acts are now in force, and there are well over a hundred separate orders under these Acts relating to counties and county boroughs. In the words of the preamble to the Royal Society for the Protection of Birds Draft Bill, 'this mass of legislation bewilders the public, the courts, and the police'.

The present system consists of protecting birds by a white list. This white list comprises the English and local dialect names of all birds which are protected; and birds not mentioned in this list can be said to be unprotected. I need not dwell on the obvious difficulties and unconscious cruelties of this system; and it is surprising that it has not long since been replaced by the black-list system. The R.S.P.B. suggests that there should be two short black lists, first of those birds which can be killed for food but which need protection during

15. Meathop Moss (Society for the Promotion of Nature Reserves)
16. Walney Island (private)
17. Calf of Man (N.T.)
18. Llanddwyn Island (R.S.P.B.
19. Puffin Island (private)
20. Askham Bog (Yorkshire Naturalists' Trust)
21. Eastwood (R.S.P.B.)
22. Cotterill Clough and Marbury Reedbed (S.P.N.R.)
23. Ainsdale (private)
24. Hawksmoor (N.T.)
25. Scolt Head (Norfolk Naturalists' Trust)
26. Blakeney Point (N.N.T. and N.T.)
27. Cley and Salthouse (N.N.T.)
28. Hickling Broad and Horsey Mere (N.N.T. and N.N.T.)
29. Weeting (N.N.T.)
30. Woodwalton Fen (S.P.N.R.)
31. Wicken Fen (N.T.)
32. Walberswick (private)
33. Minsmere (R.S.P.B.)
34. Havergate (R.S.P.B.)
35. Whipsnade Park (Zoological Society of London)
36. Dancer's End and Tring (S.P.N.R.)
37. Wytham Wood (private)
38. Bagley Wood (private)
39. Eton College (private)
40. Royal Parks
41. Belfairs Great Wood (Southend B.C.)
42. Selsdon Wood (N.T.)
43. Dungeness–Dengemarsh (R.S.P.B.)
44. Winchester College
45. Studland and Poole
46. Abbotsbury (private)
47. Brean Down (R.S.P.B. and Weston-s.-M.)
48. Walmsley Sanctuary (Cornwall Bird-watching and Preservation Society)
49. Lundy (private)
50. Llangorse (private)
51. Skokholm (West Wales Field Society)
52. Skomer (private)
53. Grassholm (R.S.P.B. and W.W.F.S.)

1. Hermaness (Royal Society for the Protection of Birds)
2. Fetlar, Hascosay and Mid-Yell (R.S.P.B.)
3. Noss (R.S.P.B.)
4. Eynhallow (private)
5. Isle Ristol (private)
6. St Kilda (private)
7. Newburgh (private)
8. Tents Muir (private)
9. Loch Leven (private)
10. Lady Isle (Scottish S.P.B.)
11. Ailsa Craig (private)
12. Scaur Rocks (private)
13. Farne Islands (National Trust)
14. Glencoyne (N.T.)

FIG. 26. Some of Britain's sanctuaries. Not all sanctuaries are marked on this map, and not all those marked are necessarily open to the public. Bodies mentioned are administrators, not necessarily owners.

the breeding season and until the young can fly, and secondly of those birds which do not need any protection at all. All other wild birds can be protected throughout the year. It seemed likely, before the war broke out, that a Bill of this kind would be put before Parliament; and it is the earnest hope of most bird-watchers that there should be no further delay. The interests of science can be preserved by reference to a suitable committee of a Ministry; and the administration of licences to collect skins and eggs should provide no great difficulty.

Meanwhile there are signs that the Government is seriously considering a Bill on these lines. If such a Bill is passed there will, of course, be many difficult administrative problems, and a great need for up-to-date knowledge and intelligence about the distribution and habits of birds. Fortunately we now have an official Nature Conservancy, and with close co-operation between it, the R.S.P.B., fact-finding bodies such as the British Trust for Ornithology and the Edward Grey Institute, and all people of good-will, we should have nothing to fear for the wonderful birds of Britain.

CHAPTER VII

Disappearance and Defiance

❧

The note of the White-throat, which is continually repeated, and often attended with odd gesticulations on the wing, is harsh and displeasing. These birds seem of a pugnacious disposition; for they sing with an erected crest and attitudes of rivalry and defiance; are shy and wild in breeding-time, avoiding neighbourhoods, and haunting lonely lanes and commons, nay, even the very tops of the Sussex Downs, where there are bushes and covert.

GILBERT WHITE, 2 September 1774

❧

THE frontiers where the world of birds impinges on that of man are largely those of emotion and symbolism. The ordinary child or man with a normal interest in birds thinks first of them as singing, or building nests, or chasing each other, or flying in excited flocks – as doing, in fact, the things which are the most highly symbolic or stereotyped of all their habits.

Many of man's habits are highly symbolic, and though born no doubt for some useful function, have, in the course of time, lost much of their immediate use. Yet they are retained, since they fulfil our emotional needs. Such habits are singing, dancing, telling stories, making puns and jokes, a lot of smoking, the curious, interesting, and rather pleasant behaviour known as 'manners', and much religious behaviour. Most of these habits are instinctive (a word which must be used with care) and few are intelligent – though that need be no reason for deserting them.

Birds have no intelligence to speak of – in the way in which we popularly understand the word. Their emotional, instinctive, and symbolic actions are correspondingly all the more important. In many ways their very emotionalism is robot-like. Their world is so narrow that normal life presents them with a very limited number of situations – and the number of possible responses they can make is just as limited. It is not surprising, therefore, that when presented with a situation quite unknown in their normal life, birds should react by doing something which does not fit (but which is itself normally a reaction to something familiar to them). Thus a gunshot may startle birds in a wood into full song; a man, suddenly intruding, may find Adelie penguins offering him stones, nesting Stanley cranes offering him sticks, a bateleur eagle in the Zoo doing his courtship display; pheasants in captivity may court their food-trough or their keepers. If a man had a bird's mind he might quite well be trusted to recite 'The Boy stood on the Burning Deck' (or some other more suitable epic) at a smoking concert, but might also recite it in an air-raid or an earthquake.

But within the web of the bird's own ordinary life its limited repertoire of actions and reactions suits it very well; and as so many of its habits of life are symbolic, like song or courtship, and for that reason curious and interesting to us, we must make some effort to examine them.

The symbolic actions of birds – those actions designed to appeal or signal to other animals in their immediate world (whether birds of their own kind, other birds, mammals, including man, or any other sorts of animals) – may involve appeals to sight, to sound, to touch, but not much to smell (birds have very little sense of smell). When these actions involve some sort of posturing, in which colour, shape, and adornment may be used, they are often called displays.

Concealment. – One of the most interesting ways in which birds 'signal' to other animals is a purely negative one – in a sense it is the opposite of signalling. It is the habit of concealment. At different stages of their lives, many different

Fɪɢ. 27. Eggs and young of ringed plover. Drawn by Hugh B. Cott, from his book *Adaptive Coloration in Animals*. Both of these are remarkable examples of concealing coloration. Note how the outline of the young birds is broken up by the narrow bands of black against the white areas, and that such daring contrasts have the effect of making the birds less noticeable than more conspicuous

Hugh B. Cott.

sorts of birds have concealing colouration, which enables them to disappear into their immediate environment and thus hide from their enemies (as far as I know there are no kinds of birds, except one or two arctic species, which blend with their surroundings so as to approach their *prey* unsuspected). Many British birds show concealing colouration or shape. Examples are:

Eggs. – Those of the ringed plover, laid on stony ground, resemble coloured pebbles, and are very hard to see (see Fig. 27). The eggs of the lapwing (another ground-nester) have a camouflage pattern which breaks up their outline. Many other waders and gulls and terns have eggs with similar protective coloration.

Nests. – Most land birds hide their nests, many by concealing them in thick bushes, in leafy trees, in long grass, and so on. But some go further and build their nests to resemble some of the special, rather than the general, features of the environment. Thus a wren may cover its nest with moss and lichens to resemble the mossy stump in which it is placed, and the neat, mossy nest of the chaffinch is well concealed in the fork of a tree. Some young birds in their nests complete an illusion; thus the young of the Ceylon black-backed pied shrike look like a branch-stump (Fig. 28).

Young Birds. – Many fledglings, particularly those which leave the nest as soon as they are hatched, have protective coloration consisting often of a disruptive pattern which breaks up their outline. Newly hatched oyster-catchers, plovers (Fig. 27), and terns are almost invisible.

Adults. – Partridges, hen pheasants, woodcock, snipe, bitterns, nightjars (by day), and several other birds have concealing coloration by means of which they blend with the general surroundings. This property is chiefly brought into use when the bird is on the nest. The poor-me-one on its nest (Fig. 29) looks as much like a branch-stump as the young of the black-backed pied shrike in theirs.

Bluff. – Many weak and harmless birds defend themselves against possible enemies by means of bluffing. I remember,

FIG. 28. Nest, eggs, young, and adult, of the Ceylon black-backed pied shrike. (After W. W. A. Phillips.)

A. The nest in position on a tree-stump.
B. The nest from above, showing the eggs.
C. Well-grown young in the nest, with the adult bird standing by.

As can be seen, the adult of this interesting shrike is far from being protectively coloured. The nest, however, is designed to resemble a mossy branch-stump, and does so remarkably closely; and when the young are still in it they complete the illusion, aided by their patchy and moss-like colouration and by the attitude all take together, facing inwards, with their tails often slightly over the edge of the nest. Unless they are being fed by their parents, the young always remain absolutely still and snag-like.

FIG. 29. The poor-me-one on its nest.

Figs. 27 and 28 have shown us eggs and young birds with a resemblance to the general features of the environment, and young birds with a resemblance to a special feature of the environment (a branch-stump). Here is an adult bird which, on its nest, resembles a special feature of its environment (again, a branch-stump). The poor-me-one is related to the nightjars, and resembles its surroundings in an amazing way owing to its concealing and outline-breaking colouration, and to its general shape. (From *Adaptive Coloration in Animals*, drawn by Hugh B. Cott.)

when I was at school, my reaction at finding a snake in a nest-box where I was told a wryneck had laid. The snake was, of course, the wryneck – stretching out its neck and hissing as hard as it could. Tits and owls will hiss in an intimidating fashion when their nests are disturbed.

Mimicry. – Practically throughout the animal kingdom we find examples of the mimicking of distasteful or pugnacious animals by harmless ones. As long as the mimics are fewer than the models, this works to their advantage, since predators, when sampling the group, are more likely to capture a model and be thus disinclined to repeat the performance. In Britain we have no good examples of birds with these characteristics, though in Madagascar we find harmless thrush-like birds mimicking pugnacious shrikes, and in the East Indies orioles mimicking friar-birds. It is possible (though not very likely) that the cuckoo's superficial resemblance to a hawk may intimidate the birds whose nests it intends to parasitize.

Intimidation. – Charles Darwin thought that bright and startling colours in male birds were adaptations for attracting the female. He thought that they had evolved by sexual selection, that is, that the choice of mate rested with the females, and that those males which were most attractive to them would be most likely to get mates and perpetuate their characters through their offspring. Practically all Darwin's important conclusions have been proved to be correct, but not this one. It is highly probable that a great element of sexual selection is found in bird (as in most animal) evolution; but it is equally probable that the bright colours and adornments of certain male birds have as their primary biological purpose intimidation and threat rather than attraction. This applies to birds which have bright plumage only in the male (like pheasants and some buntings), only in the female (like phalaropes), in both sexes all the year round (like robins and jays), or in both sexes in the breeding season (like black-headed gulls). Examples of the use of colour and adornment for the purposes of intimidation are:

Brightly Coloured Males. – On the strict Darwinian view it would appear likely that the summer plumage of male ruffs was a simple product of sexual selection. The ruff is one of the most curious waders; in the winter it is of the same build as, and not unlike, a redshank or a Bartram's sandpiper, but in the summer the males develop tremendous ruffs of feathers on the head and nape, which fold over the neck, and which, in full display (Fig. 30C), are erected to form a circle of bright colour surrounding the head. From March to May anybody can see male ruffs in full plumage and display in the Waders' Aviary at the Zoo, and prove for themselves the undoubted fact that no two ruffs are alike. I know of no bird, and certainly no wader, in which such individual variation of male aggressive plumage exists. For the function of the ruffs is intimidation rather than attraction. In the breeding season ruffs congregate at special meeting-places, usually called ruff 'hills'. On these hills the males display and threaten, with fully expanded ruffs and ear tufts, not at the females so much as at each other. In fact, the presence of the females at the display is not even necessary. Nevertheless, this intimidation ceremony is certainly sexual. Without it, the breeding-cycle of the male ruff would probably never reach mating and nesting.

At the ruff hill there is a singular absence of real combat. Battle is much too symbolized to allow much loss of blood or feathers. The males crouch and face each other, beak to beak and with ruffs spread, or chase each other to and from the little areas of the hill that each seems to call his own, but they never seem to do very much more. It appears that bluff has been largely substituted for beak-and-claw fighting, a bluff accentuated and assisted by the brightness and extent of the ruffs. It seems that ruffs 'know' when another ruff has a better ruff than they have, and in almost every case victory, in the form of undisputed ownership of the pitch, goes to that ruff which has the brightest plumage (to the human eye as well as, it must be supposed, the bird's eye).

Black game breed in certain parts of Britain (unlike ruffs,

which are known to have bred only once since 1918). Like those astonishing waders, they have meeting-grounds (in this case called *leks*) on which they congregate.[1] At these meeting-grounds male blackcock have three kinds of display – first, crowing and jumping, which appear to be a mutually stimulating affair; secondly, the highly aggressive 'rookooing', in which, to use Lack's description, 'the male stands with head and neck thrust forward, the neck enormously swelled, and utters a musical, bubbling, dove-like "rookooing", often in long continuous phrases, the whole body shaking with the calls', and in which the lyre-shaped tail is spread, the wings are dropped to show patches of white, and the red head-combs fully distended – as they are in all forms of the blackcock's display; thirdly, the circling, crouching courtship which often ends in copulation with the female.

At blackcock leks the males all hold small territories to themselves; the object seems to be to prevent other birds from interfering with the male when he is copulating with the female. The whole system of the blackcock's sexual behaviour centres round this territory; were it not efficiently secured and defended, he would have little sexual success. Hence the importance of his bright plumage, his use of it in display, and his song (for this is what 'rookooing' is). That his plumage is used in a socially stimulative display, and that it is used in a courting display directed towards the female, once she has walked into his territory, seems to be a secondary matter – the important fact is, that, by intimidation and bluff,[2] he has carved out a territory for her to walk into.

The snow-bunting is a bird which nests sparsely in the

1. The reader is referred to two very important papers: E. Selous, in *The Zoologist* for 1909 and 1910, and David Lack, in *British Birds* (1939), vol. 32, pp. 290–303.
2. The blackcock's intimidation display is not all bluff; sometimes mortal combats take place. This is wasteful, of course, in the biological sense, but seems to be rare enough.

north of Scotland, and widely in the arctic.[1] It is a monogamous bird (though occasional polygamy has been noted),
and the male in breeding plumage, with his attractive black
and white colouration, contrasts strikingly with the brown
and white female. Though his contrasting colouration is
brought into use in special display ceremonies devoted only
to his female, yet its primary function appears to be one of
threat (see Fig. 30A). When the male has arrived in his breeding territory in spring and notices another snow-bunting,
whether male or female, approaching his territory, he goes
at once into a threat attitude. He faces the intruder, lowers
his head between his shoulders, and utters a sound which
Tinbergen writes as *p*EEE, and likens to the sawing note of
the coal-tit. If the newcomer alights in his territory he flies
towards it, singing in flight, and rising steeply with his body
curved upwards and his wings trembling. Only when he is
near the other bird does he distinguish in his reaction between male and female. If the intruder is a male, a struggle
may result; if a female (and he is unmated), a courtship
display. Here then is a case of a first general reaction being
entirely one of threat, continued as such if the other bird is
recognized as a male, and substituted by a courtship display
only when, *at close quarters*, the newcomer is recognized as a
female.

To observe ruffs it is necessary to go to Holland; blackcock leks are few and far between; snow-buntings take a lot
of finding on remote Scottish hilltops; we must therefore
find another example a little nearer home. Let us examine
the reed-bunting[2].

Eliot Howard (who will be mentioned a good deal in the

1. I am using it as an example, since its habits have been so beautifully
worked out in East Greenland by Dr N. Tinbergen (*Transactions of the
Linnaean Society of New York*, vol. 5). This tremendously interesting paper
contains much valuable discussion on bird behaviour, particularly on
the theory of bird territory.

2. The breeding behaviour of the reed-bunting is beautifully described in Chapter I of Eliot Howard's *An Introduction to the Study of
Bird Behaviour* (Cambridge, 1929).

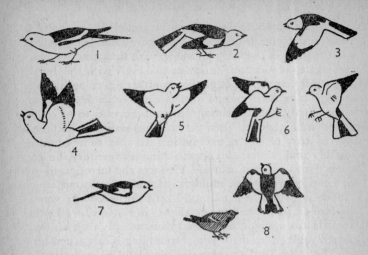

A. Some displays of the male snow-bunting, after Tinbergen:
  1, 2. Threat.
  3, 4, 5. Song-flight.
  6. Territorial fight.
  7. Calling on arriving in the territory.
  8. Displaying to newly arrived female.

B. Intimidative attacks by black-headed gulls at stuffed mounts (winter and summer plumage) set up in or near their territories.

FIG. 30. Various types of bird display (pp. 130-3).

c. Aggressive postures of the ruff.

D. The male great bustard in full display. The bird is viewed from the side, and its head is to the right. Highly contrasting feathers are erected in many different directions, and the throat is inflated until it becomes the immense balloon-like bag that hangs down almost touching the ground in front. The head itself is concealed among the feathers.

E. The intimidative displays of pheasants are sometimes quite fantastic. Probably the most far-fetched is that of Bulwer's pheasant. The hammerlike appearance of the head is caused by huge, blue wattles, which, in display, are fully inflated. The tail is white, and spread sideways towards the object of the display. (After Heinroth.)

F. Back view of the greater bird of paradise in full display. Note the butterfly-like arching of the wings, and the way the plumes of the back are erected. The bird may quiver in this position for many minutes, often sidling up and down the branch, and may even turn upside down in the height of its emotion and continue its display hanging from its feet.

G. In display the sun-bittern exhibits the contrasting coloration of its wings and tail and looks like a great butterfly. This display may mean rivalry when directed towards another male, but also appears to have a function of mutual stimulation when made between a male and female.

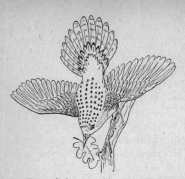

H. Part of the courtship display of the grasshopper warbler consists in the offering of a leaf (a probable symbol of nest-material) with quivering wings and expanded tail. (After Howard and Grönvold.)

I. The mutual courtship of the kagu (sometimes to be seen in the London Zoo) may produce postures such as this: male and female face each other with fully erected crests.

J. Fulmars have greeting ceremonies when they meet their mates on returning to the nest-site (or when they visit the nest-sites of other pairs). Often these ceremonies consist of bowing, as in the drawing, or of beak-scissoring.

K. The greeting ceremony of the kittiwake consists mainly of mutual loud 'kitti-waa-aaking'.

next chapter) has given us a picture of its behaviour whose accuracy of description is equalled only by its quality of prose. Though we cannot quite compare his account step by step with Tinbergen's description of the behaviour of the male snow-bunting, we can find in it enough to tell us that the male reed-bunting, no less than its arctic counterpart, lives a spring life, in which intimidation, aggression, and territory occupy its mind to the exclusion of almost every other instinct. Intimidation (Howard calls it excitement) is expressed by voice, flight, and movement at rest. Song, the chief signal of the possession of the territory, doubles in its intensity at the approach of another bird. Sexual flight seems to be of two sorts; one like a butterfly – slow flaps with full wings; the other like a moth – rapid vibrations of partly opened wings (compare the threatening flight of the snow-bunting). Perching on a branch or the ground, the male flutters one or both wings, or expands his tail. These displays accompany attack on a territorial rival, but when the female is at close quarters they are transferred into what a dancer might call the 'routines' of a chase that may end in successful or unsuccessful mating. We cannot go very far wrong if we interpret his general reaction as being one of threat (like the snow-bunting) and his special reaction battle or courtship, depending whether, at close quarters, he finds a female or another male.

Brightly Coloured Females. – Some birds, like phalaropes and button-quails, have brightly coloured females and relatively dull-coloured males. The rôles of male and female in the life-cycle seem to be completely reversed; when the female has laid the eggs it is the male which has to sit on them. It is not surprising, therefore, to find that, among these peculiar birds, the female takes the initiative in marking out the territory, intimidates other females, and uses her bright colouration in male-like postures.

Bright Colours in Both Sexes. – If it is true (as we shall see later) that the sexes are often recognized by their posture and movements, even among some birds in which the sexes have different colours and adornments, how much more true must it be in many cases where male and female have almost identical plumage. In the breeding season the male goldcrest has a more golden crest than the female; otherwise they are outwardly identical. In an encounter between rival males the crests are erected and used as the special signalling organs of threat attitudes. They are used in courtship also, but in this case the erection of the crest by the male evokes a different response from the female. Much the same situation can be seen when watching the use of their contrasting colouration by great tits or blue tits in sexual rivalry or courtship.

Robins,[1] as David Lack mildly puts it, are 'renowned for their pugnacity'. Without any doubt they are the most highly aggressive of any British bird. Intimidation is their daily life. Bright colours, with an *entirely* aggressive purpose, are found equally in both sexes and, except in part of their breeding behaviour, so are intimidation and aggression. In autumn and early winter both sexes hold territories which all except a few of the females defend by song and by aggressive posturing. The male continues his territorial and aggressive activities practically throughout the year, being joined in the spring often by a female from a neighbouring territory, who then ceases to be a rival and becomes a mate, ceasing also her song and aggressive display.

The robin's organ of aggression is its red breast. It is used solely for this purpose. It is not even used (as are aggressive organs in most other birds) in courtship with the female at close quarters; the only form of robin courtship seems to be the feeding of the female by the male.

Robins posture by stretching throat and breast to display the maximum amount of red, and accompany this by sway-

1. The most intensive study on the life-history, aggressive and recognitional behaviour of any British bird, is that of Mr David Lack on the robin. The bulk of it is published in the *Proceedings of the Zoological Society of London* (Series A), vol. 109 (1939), pp. 169–219, and in the *Life of the Robin* (London, 1943, new ed. 1946).

ing from side to side. Colour, posture, and song are important, since robins attack the red breasts of whole stuffed adults, just the red breast alone stuck up on a piece of wire (in one case, after many attacks, a mount was taken away and the robin continued to attack the place in the air where it had been), or even a stuffed bird with the red of the breast covered with brown ink; robins also pursue other robins (or even other species of birds) when they see them flying away; they also reply to song by song, or to a distant view of another robin by song. Their mates alone, whom they recognize individually, are immune from this sort of treatment.

Though highly aggressive and pugnacious, the robin's behaviour can reasonably be regarded as being mostly bluff. A cage with wild robins in it will not be included in the territory of a robin outside, though the occupants of the cage can only intimidate the intruder through the wire-netting.

Bright Colours in both Sexes in the Breeding Season. – Many kinds of birds have a seasonal change which is of the same nature in both sexes. In winter the black-headed gull has a white head (with small dark markings) and in summer a rich chocolate-brown one in both sexes (the head is never black). It seems that this donning of summer plumage in both sexes is found very widely in social birds, like some gulls, guillemots, and puffins; less often in non-social kinds.

In a black-headed gull colony both birds of the pair take a roughly equal part in all the business of breeding, and both defend, by means of threat, a small area round their nest. Hostility will be shown by a nesting gull of either sex only when the territory is invaded – gulls will tolerate each other in the neutral ground away from the nests. By using stuffed 'mounts' we can find what it is in the intruder that causes the hostility (Fig. 30B). A stuffed black-headed gull in full breeding plumage, if put in a territory, is vigorously attacked by the owners, not only with their beaks (with

1. Anybody who is interested in black-headed gulls, their social lives, territory, breeding, instinct, and intelligence, cannot do better then read F. B. Kirkman's *Bird Behaviour* (London, 1937).

which they attack live intruding gulls) but also with their feet. Normally, black-headed gulls attack only human intruders with their feet – it can be imagined that the stuffed mount is the only sort of invader likely to wait until the defender can get its feet into action.

Most of these attacks are directed against the 'black' head of the mount; clearly this adornment is important. We can find, by experiment, the limits of its importance. If we set up another stuffed gull, this time in winter plumage – with a white, not chocolate, head – we see that it is attacked almost as vigorously as the mount in full breeding plumage. If we set up a corpse of a gull it will be attacked as long as its pose bears some relation to that of a living bird. If we destroy the corpse's realism by cutting off its head, for instance, it will no longer be attacked, though the gulls may be apparently puzzled and frightened by it.

So we have established that the black head is not all-important in provoking intimidation, but that pose is important too. But the black head seems to focus the display; and, even by itself, it can 'release' the aggressive display of the territory owner, provided it is in the right position. Set up on a skewer at normal height, the head of a decapitated bird is attacked with beak and feet. If it falls to the ground, on the other hand, it seems to be no longer capable of releasing aggressive behaviour; its position is now more that of an egg than of an intruder. In fact, sometimes when this happens the living bird may roll the head into its nest and attempt to incubate it.

Submission. – The complicated system of postures, movements, and the use of coloration and song represented by aggressive display has a definite biological purpose. It is of use. In their sexual rivalry, birds seldom fight to the death; rather do they play a game of chess in which their territory is often the chess-board, and their postures, colours, and adornments the pieces. The competitor or intruder nearly always 'knows when it is beaten' before feathers or even tempers are lost. Clearly this system is a highly economical one – the loser is unhurt and free to carve out a territory elsewhere, or to try his powers of bluff on a different male. The species as a whole is prevented from suffering from the waste of its precious individuals. As we will discuss in the next chapter, the territorial system is of such usefulness and importance that birds as a whole can afford this complicated system of sorting themselves out in the breeding season.

In certain kinds of birds it seems to be necessary to have additional devices to avoid actual battle during or after a display of intimidation. This is particularly true of social birds. Among birds which normally live in flocks, and particularly among those in which flocking extends into or even through the breeding season, the normal releasing of sexual rivalry might lead to serious complications were not some such system in existence. Thus we find that jackdaws, when threatened by others of their kind, may turn and expose the least protected part of their bodies – the backs of their necks. In the jackdaw this is even specially marked by a grey patch. When this submissive action is performed, the aggressor is assured of moral victory, which in a sense is all he wants. Submissive attitudes have been recorded in several other social birds (the night-heron, for example), and, now that their importance is realized, they are being found to exist quite widely.

The importance of submission is shown in an interesting way by the behaviour in close confinement of such male birds as do not submit. (In nature these move off somewhere else after an exchange, which they cannot do in captivity.) When several birds are confined together, one soon becomes the despot, and when he is removed, another dictator takes his place. In fact, such birds can be arranged in descending order, according to which pecks which. Four male blackbirds, which Tinbergen kept in a small cage, 'developed a severe despotism, as captive birds are apt to do in the spring, one of the males actually killing all the others.' This certainly would not have happened had blackbirds been naturally social birds, for then it is probable that some submissive device would have been developed to prevent such mortal combats. It is interesting to note that the domestic fowl, which in nature is not a social bird, has in the course of some thousands of generations of captivity developed a system by which it can hold a painless social hierarchy, and that submissiveness plays a large part in preserving peace in the farmyard. That this does not always apply can, however, be deduced from the continued existence of fighting cocks.

CHAPTER VIII

How Birds Recognize One Another

❧

I have now, past dispute, made out three distinct species of the willow-wrens (Motacillae trochili) which constantly and invariably use distinct notes. But at the same time I am obliged to confess that I know nothing of your willow-lark. In my letter of 18 April, I had told you peremptorily that I knew your willow-lark, but had not seen it then; but when I came to procure it, it proved in all respects a very Motacilla trochilus, *only that it is a size larger than the other two, and the yellow-green of the whole upper part of the body is more vivid, and the belly of a clearer white. I have specimens of the three sorts now lying before me, and can discern that there are three graduations of sizes and that the least[1] has black legs, and the other two[2] flesh-coloured ones. The yellowest bird is considerably the largest, and has its quill feathers and secondary feathers tipped with white, which the others have not. This last[3] haunts only the tops of trees in high beechen woods, and makes a sibilous grasshopper-like noise, now and then, at short intervals, shivering a litile with its wings when it sings, and is, I make no doubt now, the* Regulus non cristatus *of Ray, which he says 'cantat voce stridulâ locustae'. Yet this great ornithologist never suspected that there were three species.*

GILBERT WHITE, 17 August 1768

❧

WE must avoid placing ourselves in the birds' position when we deal with the methods by which they recognize one another, that is, their own or the opposite sex, their parents or offspring, or their species. We have no philosophical justification for imagining that, because we recognize differences between a male and a female bird or between two different sorts, the birds themselves recognize these differences by the

1. The chiffchaff.   3. The wood-warbler.
2. The willow- and wood-warblers.

same characteristics. Only by a long process of objective observation, by taking nothing for granted, and often by the help of actual experiments in the field, can we find out the real basis of recognition in birds. It is becoming clear, as I have already hinted, that in many cases postures and movements are just as important as, or even more important than, special colouration in sex recognition. The male flicker (a kind of American woodpecker) is distinguishable to man from the female, since he possesses a handsome 'moustache' of black feathers on each side of his chin. Apparently a flicker can also distinguish a male from a female by this characteristic, as has been proved by some recent field experiments. A pair of birds had settled down to normal breeding behaviour, and were clearly recognizing each other by habit or by posture and movement, or by a combination of both. The female was then provided with a false moustache. When she was released the male approached her from behind and began to mount her (the pair had reached this stage in their breeding cycle). She turned round and he saw her moustache. Immediately he went into his full (anti-male) aggressive display, pursuing the unfortunate female for two and a half hours and spreading his tail to show the bright undersurface.

This experiment shows, first, that flickers recognize their sex by posture, secondly, that the moustache, an adornment, is also a recognition mark, thirdly, that the contradiction artificially produced between male (moustache) and female (posture) characters produces great excitability and a generally anti-male reaction, and finally that the exhibition of the bright under-surface of the tail during the chase may be (in its colour and the way it is shown) a further male-recognitional character.

In the snow-bunting Dr Tinbergen found that it was mainly the attitude and movements of the bird against which threat display was directed that determined whether the threat should be continued as such or transformed into courtship display. This explains some of the occasional cases of apparent homosexualism in birds. If, by some upset of its sex-glands, or another cause, a male bird goes into a female attitude, it will often be accepted as a female, even though it may bear male plumage – the displaying male may even try to mate with it. Among the ruffs and reeves in the Waders' Aviary at the Zoo there are many more males than females, and some of these often take up female attitudes. The consequent attempts at mating have often led onlookers to suppose that reeves grow ruffs.

Observers of the sexual behaviour of birds in the field have often (as described in the case of the black-headed gull) used stuffed birds to test reactions and recognitions. In some cases this has been the means of discovering how the nature of a bird's sex is recognized by a rival or by a possible mate. If a male ruffed grouse (an American bird) is presented with the skin of a male of its own species stuffed in an ordinary attitude, it will try to copulate with it – showing that in this particular case it is the attitude, rather than the characters or adornments, that is important in recognition, since the attitude in which a bird is normally stuffed is essentially female, having, as Dr Tinbergen points out, the lack of movement and rather hunched position of a female willing to copulate.

The blackcock does not show the same reactions to stuffed mounts as does the ruffed grouse. This bird (in which the sexes are more widely different) always recognizes the sex of the dummies; males copulate with stuffed females, strike at stuffed males in the normal position, and, with a male stuffed in the 'rookooing' attitude, may strike repeatedly at the adornments of the head. One bird eventually removed the head and continued to strike at it where it lay three feet away.

We have now discussed two mechanisms, important in the recognition of the sexes – colour (and adornment) and posture. Male birds may distinguish males from females by one or both of these characteristics. There remains another important mechanism used for the purpose of recognition –

voice and song. Vocal sounds play their part in enabling
birds to recognize one another individually.[1] Gentoo penguins recognize their mates by crowing, though they sometimes make mistakes.

Song is a very important factor in enabling certain species
which are very similar in plumage, and which overlap in
their breeding range to distinguish each other. In Britain
we have two examples of pairs of species which are almost
indistinguishable in size, colour, and plumage: the chiffchaff
and willow-warbler, and the marsh- and willow-tits. Both
these pairs of birds are very closely related, by the evidence
of their anatomical structure, and it can be imagined that it
is not very long since, in the evolutionary perspective, they
have each shared a common ancestor.

Without going into the arguments about how these species
of birds have come to share the same geographical distribution, we can examine the devices by which they are able to
distinguish each other, and so keep their identity as entirely
different species. In both cases, warblers and tits, the important mechanism seems to be song, or at least voice. The difference between the song of the chiffchaff (harsh, loud chirp)
and that of the willow-warbler (sweet, plaintive note (Gilbert White)) is tremendous, so much so that it is the chief
character which enables the human being, investigating the
bird life of woodland, garden, or scrubland, to tell them
apart. Though their habitats are not quite the same – chiffchaffs like close woodland with high song-posts, while

1. Recognition, by birds, of individuals of their own species is a fact,
though not a universal one. A lot of work is being done on this subject;
among recent discoveries are: black-headed gulls and song-sparrows
recognize their neighbours, but not strangers; paired willets (an American
wader) can recognize each other without a display; gulls and penguins
recognize and greet their mates from afar; immature night-herons
(paired) recognize each other after 20 days' separation, but not after 6
days if extra feathers are glued to the head of one; young night-herons
do not recognize their parents, but herring-gulls recognize them, and
their brothers and sisters as well; most social birds recognize their own
young.

willow-warblers seem to prefer open scrubland and can tolerate low song-posts – there can be no doubt that it is the difference in song which prevents the birds themselves from making a serious mistake in identification. The wood-warbler, it might also be said, is outwardly very similar to those two, and also has a very distinctive song (sibilous, shivering noise (White)).

Like these warblers, the marsh- and willow-tits overlap in broad geographical distribution, and a good deal in habitat. Although they are superficially so much alike that ornithologists did not separate the British species until 1900 (the chief anatomical difference between them lies in the structure of the feathers of the head), yet they are able to distinguish their own species, and thus avoid wasteful attempts to interbreed, by their call notes and songs. These are absolutely distinct. The call notes[1] of the marsh-tit are four, which can be written, *pitchuu*, *tsee-tsee-tsee*, *chick-a-bee-bee-bee*, and a harsh note *chay*. The last is the only one which is at all like the *chay-chay-chay* of the willow-tit, which in any case is much more harsh and twanging. The willow-tit never uses *pitchuu*, though it may make a sound *chich* in its place, and its thin *eez-eez-eez* or high *zi-zit* are not found in the marsh-tit at all. The songs again are entirely different in the two species – the *piu-piu-piu* of the willow-tit is quite unlike the *schuppi*, *schip*, or *pitchaweeo* syllables of the marsh-tit.

Practically all the finest songs of British birds are of those whose sexes are alike (the blackbird and the blackcap are exceptions, but it is possible to regard these birds as being in the course of evolution towards a state in which the sexes will become of similar plumage). Though the main function of song in most perching birds must be regarded as signalling the possession of a territory, we are prompted to believe that the quality, extent, and varied nature of these

1. *The Handbook of British Birds* (London, 1938), vol. 1, pp. 263, 266. The descriptions of bird voice in this book, the foremost national bird fauna in any language, go about as far as possible in rendering bird sounds into some recognizable sequence of consonants and vowels.

FIG. 31. The overlap in Britain between the hooded and carrion-crows. The hooded crow is the crow of Ireland, the Isle of Man and north-west Scotland; the carrion-crow that of England and Wales. Over the intervening areas in Scotland the two birds overlap, and, to a slight extent, interbreed. The area of overlap is double-hatched. The broken line encloses areas in which breeding carrion-crows are occasional and irregular, though increasing ; the dotted and broken-line areas formerly occupied by some breeding hooded crows, but from which they are now quite absent.

songs also form a mechanism by which the sexes recognize each other. With birds like nightingales and warblers, it is reasonable to suppose that a very distinctive song, in the absence of distinctive and recognizable colour, is a great help to a female in finding a mate, and distinguishing him when he is found.

Occasionally, in the course of evolution, a situation is produced in which such distinguishing characters as have been developed are not quite enough to keep the species entirely apart. Such a position exists with regard to the carrion-and hooded crows. Derived originally from a common ancestor, these birds have probably enjoyed, in the past, complete geographical separation, and during the period in which they lived apart they diverged from one another, one becoming (or staying) black and the other developing its typical pied plumage. Later in time, the hooded crow has extended its bounds until it has once more come into contact with its cousin, the carrion-crow. In Britain the line of breeding overlap (Fig. 31), which appears to be slowly moving north-west at the expense of the hooded crow, extends from Ireland across the lowlands and part of the highlands of Scotland. Here the two birds breed in the same area, and, although for the most part they keep themselves to themselves, a remarkable number of cases of interbreeding (with fertile offspring, as far as can be seen) have been recorded. In the case of the crows, then, it is clear that one apparently distinguishing characteristic is not completely effective in preventing crossing.

CHAPTER IX

Territory – A Theoretical Case

❦

*During the amorous season, such a jealousy prevails between the male
birds that they can hardly bear to be together in the same hedge or field.*
GILBERT WHITE, 8 February 1772

❦

The General Statement. – For the last seven or eight years bird
literature, scientific, aesthetic, and popular alike, has been
full of discussions of the territory theory. It would only be
going over old ground to put this important contribution towards our understanding of birds in any sort of historical
perspective; it should be enough to say that the idea that
birds defended plots of ground in the breeding season against
members of their own species was realized in 1622, restated
in 1632, expanded in 1772 and 1774 (by Gilbert White and
Oliver Goldsmith), and occupied a treatise in 1868; the plot
of ground was christened the 'territory' in 1903;[1] the modern
territory theory was first properly stated in 1907,[2] expanded
to book form in 1920,[3] recapitulated in 1929,[4] reviewed in
1933,[5] and has been the subject of heavy criticism (comment
rather than derogatory quibbling), investigation, and
experiment ever since.

There are two modern definitions of bird territory which
it will be well to keep in mind. A general one (Tinbergen)

1. C. B. Moffat, in the *Irish Naturalist*, pp. 152–66.
2. H. Eliot Howard's *The British Warblers, etc.* (London, 1907–14.)
3. H. Eliot Howard's *Territory in Bird Life* (London).
4. H. Eliot Howard's *Introduction to the Study of Bird Behaviour* (Cambridge).
5. David and Lambert Lack in *British Birds*, vol. 27, pp. 179–99.

is 'whenever sexual fighting is confined to a restricted area, that area is a territory'. A more special one (Lack) describes territory as an 'isolated area defended by one individual of a species or by a breeding pair against intruders of the same species and in which the owner of the territory makes itself conspicuous.'

We must also bear in mind that the machinery of keeping a territory is based on the aggressive coloration, adornment, and action which we have discussed in the last chapter; with this definition and condition we can attempt to describe a bird's territorial life. The bird we will select is imaginary, though there are many living birds like it; imaginary because no one species of bird shows all the typical attributes of a territory-holder, and because it will be to our advantage to discuss the rules before the exceptions, the type before the examples.

Our standard bird, then, is a land bird – a passerine (or perching bird); inhabits hedgerows, gardens, scrubland, or woodland; has a brightly coloured male and a dull female; has a male which sings; is partly migratory and partly resident; has catholic tastes in food; is, on the whole, monogamous. In winter mixed flocks of males and females work the hedgerows, fields, gardens, wood-edges, rickyards, or beaches above high-water mark of some country or countries in the Holarctic Region – temperate Europe, Asia, or North America.

At the end of the winter – perhaps in early March – these flocks cease to be the entirely amicable affairs they were since autumn. They still retain their function – that of co-operative food finding, so that the discovery of one is shared by the others; but quarrels, not over food so far as can be seen, nor over anything tangible, begin to break out between males. They may flirt their wings or posture at one another, and have little fruitless chases – nothing serious, but enough to show that some internal change is taking place in the males' make-up that prevents them, temporarily, from behaving like ordinary members of the flock.

As time goes on these quarrels tend to get slightly more serious and more prolonged, and are accompanied by little expositions of the breeding song, but before they become really acute the males begin to go away for short periods instead. They pay visits to suitable small areas of their summer breeding habitat. At first these visits are very short, and involve the prospecting of a fairly wide area of the summer habitat; as time goes on the period spent away from the flock gets longer and longer, and the area visited smaller and smaller, until a day comes (or a night, rather) when the male has left the flock altogether and sleeps in what is now the 'rough copy' of his breeding territory.

By now he is well acquainted with his home. The boundaries of it interest him little as yet; most of his business concerns several points of vantage (we can call them headquarters), to which he continually hops or flies, and from which he sings first a few quiet bars, or a little shadowy copy of what, in a day or two, he will be pouring forth with all his might as his full spring song. The paths to his headquarters become as clear to him as is, to us, the position of the handle on the bathroom door; the post or gate or branch from which he sings is approached perhaps by a flight to a bush, another to a branch, a hop to the left, a quick flight to the right, and a hop straight upwards: he seldom departs from this routine.

As time goes on he visits his headquarters more often, and may sing for as much as an hour on end; he seems to favour one or two song-posts more and more, so he cuts the number of his headquarters down, though, when passing the immediate neighbourhood of those spots which he has discarded, he may still go through his stereotyped routine of path-movement. With the increase of his song and preoccupation with headquarters he goes less far afield for foraging, until he is feeding entirely in the loose area he owns.

This area does not stay loose and undefined for long. Other males of the same species, who have also left the flocks, approach, fly past, sing, come foraging in his area. He is not tolerant of these birds. He chases them, postures aggressively at them, and sings at them. Soon he finds that certain of these birds are his neighbours, owning head-quarters on the other side of the garden, on a branch of the next big tree, or fifty yards down the hedge; towards these his aggressive reactions are more patterned; he plays them at song-tennis over the neutral ground between – ground which rapidly becomes less and less neutral. When one of these neighbours, now become rivals, alights in his area, he makes a display flight at it, often singing on the wing; the neighbour retreats into his own area, and if he is followed, the rôles are reversed; in such a way the boundaries of the territories are marked out. Eventually it becomes possible for the human observer to plot the borders of the territories on a map, almost down to the last blade of grass. Over these borders are fought the most bitter battles – here is there least to distinguish the rivals in pugnacity. For our birds, unlike some human nations, recognize the aggressor. The farther a fight ranges into the home bird's territory, the more aggressive does he become, and the less aggressive his rival; only on the borders are they equally matched – in fact, the border *is* where they are equally matched.

When our male first became resident in his territory, he owned about two acres. His battles with his neighbours have cut the area down to one; he may now be in relation to four or more other males on different sides, each with about an acre, too. It will be perhaps about three weeks since he left the flock.

It is now the females' turn to leave the flock. They seem to forage individually farther and farther afield; one passes through a male's territory and is threatened as if she was a male; her time is not quite ripe, for she takes no notice of his song and goes on her way. The next one that arrives has been attracted by the male's song – she lands in the territory and approaches. The male greets her in exactly the same way as he would a rival male entering his territory – by threat. Not until she is at close quarters does he recognize her and change his attitude. Then comes a sexual chase. Round and round the territory the two birds fly, the male trying to grasp the female's back, bring her to the ground, and mate with her. Eventually he gives up, and both land, the male often finishing with a little courtship display, and often a call that seems to represent frustration. For, though the male is ready to copulate, the female is not, and may not be for some days or even weeks – although she was ready to be attracted into the territory in the first place.

Sexual chases now become established as a sort of daily routine, and though they indicate that successful copulation has not yet taken place, they show all the same (as Dr Tinbergen points out) that the birds are paired. Though the female will not allow the male to copulate with her, she does not leave him. Apart from their frequent sexual flights, the birds learn to recognize each other individually, thus dispensing more and more with the usual recognitional displays. The female helps the male to protect the territory (and his aggressive attitude to his neighbours increases still more), not by attacking his rival males, but by attacking other females. The male's song stops almost entirely as long as his female is in his territory; if she happens to wander out of it (and she does so less and less as time goes on) he starts singing again.

Though at this stage the male's aggressive behaviour increases, it is still stronger towards the centre or headquarters of the territory, and less strong towards the borders. Even now, when neighbouring males are mated, and consequently more pugnacious, a new male, late from the flock, may (no doubt encouraged by the absence of song) try to stake out a territory between those already occupied. Sometimes he may succeed, depending on the distance between his chosen headquarters and that of the birds already in possession, for up to a certain *minimum* limit territories are compressible. If he does not transgress this limit, he may succeed; if he does, he gives up the struggle and passes on.

One day the female shows great interest in one or two bushes, or a stretch of hedge. She picks up a little piece of nest material. The male approaches her, and this time the normal sexual chase does not follow. She is ripe for copulation, the state of her breeding organs has crossed a certain threshold; with flattened back and lifted tail she invites the male, and he mounts and copulates with her. But even now this does not always happen – she may invite, but the male may not be receptive, and then she will relinquish her attitude; the male may invite, and another sexual chase take place; copulation may be unsuccessful, and may result in courtship posturing. Rapidly, however, the birds become used to one another, and the female starts building the nest in earnest. The male too shows signs of building activity; he may pass building material to the female, may add it to the nest himself, or may make feeble and unfinished 'cock-nests' of his own. From a week to a fortnight after the first copulation the nest is complete.

The female now becomes restless, and makes frequent visits to the nest. Once she visits it for over half an hour; she does this again the next day and the next and the next. Four eggs are there when we look, though during those four days of laying the nest was still being lined with feathers. The female no longer allows the male to copulate; for his part he resumes his song[1] – it is almost as strong as if he was unmated, though he has been almost silent since she arrived in his territory. A day or two after she has laid the fourth egg she begins to sit. The male takes no part in the incubation, but continues his territorial song and aggressiveness, in the intervals of foraging for himself and taking food to the female on the nest.

1. Dr Tinbergen points out that the male snow-bunting sings especially when his female is absent; that his song increases when she refuses copulation; that it almost reaches its original strength when the female begins incubating. This probably indicates that sexual potency still exists; at this period the male has a chance to be 'unfaithful', and though normally monogamous, may actually mate with a new female.

After a fortnight the eggs hatch. The young are like those of most passerine birds, at first almost naked and blind. The business of feeding them occupies the parents almost the whole of their time. For a day or two the young are fed from the crop; after this large quantities of insects are needed, and are brought to the nest three or four times an hour. As the nestlings grow larger and hungrier, every hour of daylight means five or six, then eight or nine, then twelve, and, towards the end, sixteen visits. To collect these insects the parents have to forage wherever insects are, and parties of parents will hunt together in neutral ground or even over part of the territory of one of them, and even in the latter case little aggressive behaviour will be aroused. Every day the young grow more noisy and more active. A fortnight after hatching their excrement is no longer voided in a little bag which can be removed by their parents, but then there is no further need to avoid fouling the nest; though they cannot fly they have left it. They hide now, a special call signalling their presence to their parents. Sometimes the parents each take charge of half the fledglings. After a week or ten days the young begin to feed themselves, after another week they are no longer fed by their parents at all, in another day or two their parent-calls have been replaced by the ordinary adult flocking call, and they are off on their own.

Long before they finally go altogether the young have been wandering more or less where they would. Even when fed by their parents they have trespassed into neighbouring territories, where they and often their parents too have been tolerated. But when they are finally independent their parents are found to be defending their own territory again, the male, who was too busy to sing while he was feeding the young, hard at it, the female refusing copulation until she is ripe for the recurrence of her cycle that means a second brood. And after the whole process has been gone through a second time with the second brood, the parents (unless indeed they have a third or fourth) will lose their urge to copulate (an urge which, we have seen, is of longer and steadier duration in the male than in the female); they will lose their territorial sense; flock communication sounds will replace male song; like their young, they will wander off foraging, they will moult, and the autumn flock will be established again.

CHAPTER X

Territory, Courtship, and the Breeding Cycle

❧

*Hedge-sparrows have a remarkable flirt with their wings in breeding-time.
Redbreasts sing all through the spring, summer, and autumn.*
                    GILBERT WHITE, 2 September 1774

*The missel-thrush is, while breeding, fierce and pugnacious, driving
such birds as approach its nest with great fury to a distance.*
                    GILBERT WHITE,[1] 13 September 1774

❧

THE composite bird which we have described in such detail
owes much to the buntings and to those who have so carefully studied them.[2] Now we have a picture (which I hope is
typical) of the life of a territorial bird, and we can compare
the lives of real birds with this picture. We have traced the
breeding history of our standard bird, stage by stage, from
the moment when the winter flocks break up until the
moment when the autumn birds show signs of flocking again.
On our way we have passed several milestones; let us take
the journey again now, resting for a time at each milestone,
and see what the different sorts of birds are really doing.
Our milestones are:

1. His life-histories of the house-martin, swallow, sand-martin, and
swift should be of particular interest to readers of this chapter: see his
letters to Daines Barrington of 20 November 1773, 29 January, 26 February, and 28 September 1774.
2. Besides Tinbergen's and Howard's work (already cited) the reader
is enthusiastically referred to Mrs Margaret Morse Nice's 'Studies in the
Life History of the Song Sparrow,' published in 1937 and 1943 in the
Transactions of the Linnaean Society of New York, vols. 4 and 6.

(1) *The Taking up of Territory, and Winter Territory.* – Territories can be taken up by males (most cases, like our typical buntings), by both sexes together (particularly among sea birds and some arctic waders that pair up on migration, before they have arrived at their breeding-place), or by females (for instance, phalaropes and button quails, where the female has the bright colours and the aggressive display). To take up the territories the males (we shall exclude the other cases) leave first for the north, if they are of a migratory species; or stay more or less where they are, if they are highly resident or hold winter territories. Birds which do the latter deserve rather a special mention. If we look back to pp. 147–48 we see that Tinbergen's definition of a territory hints at a sexual purpose, while that of Lack (who has extensively studied the robin, which holds a winter territory) carefully avoids doing this. If we look – in the evolutionary perspective – at the business of winter territory, in which males and females – mates in the summer – may be rivals in the autumn, we can see that these two definitions may not be contradictory, provided, as seems likely, one habit has evolved out of the other. It is possible to imagine that the keeping of winter territory (by such species as practise it) may, in the course of evolution, have been an extension or resumption of the normal breeding-cycle territory, and that though originally a purely sexual affair, it now acts in the winter for other purposes. For instance, it keeps the species evenly spread in regions of scanty food,[1] and in small units invisible to predators rather than in large visible ones.

Winter territory is being found to persist, to a small extent, in quite a number of birds. Those in whose lives it plays an important part range from California shrikes and mockingbirds to San Francisco spotted towhees; in Britain the male blackbird holds a winter territory to a certain extent; and

1. Of course the effectiveness of this depends on the kind of food. Many insect-eating birds get a lot of advantage out of winter flocking, since a group or swarm of hibernating insects, turned up by one bird, becomes the prey of all.

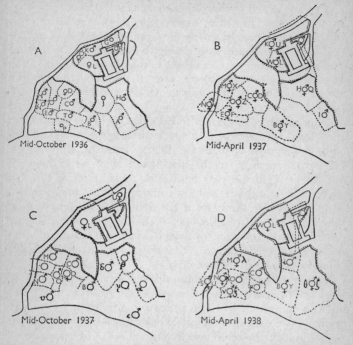

FIG. 32. A two-year history of robin territories. Each letter, Roman or Greek, stands for a different individual robin. Circles represent occupied territories, whose boundaries are the dotted lines; o – sex not known, ♀ female, ♂ male, ♂̧ pair, male's letter on left. From figures A, B, C, D, representing the situation in succeeding summers and winters, it is possible to trace the individual bird's territories, and the history of their mating in successive years. (After Lack's very detailed study.)

some females hold them also. The finest example in Britain, though, is the robin.

A robin may inhabit only a few acres during the whole of its life, though certain females wander, and a few are even migratory. In winter (contrary to the impression given by most Christmas cards) both male *and* female hold individual territories, which *together* are almost exactly the size of their summer ones. These they defend, like most aggressive birds, by song and dance. In spring the females cease their activity, and very often mate with a neighbouring male rival (thus sometimes getting the same mate in successive years) (Fig. 32). In early autumn alone, after the waning of breeding activity, is there a time-gap in which territorial activity is low.

Only one bird is known to be more highly resident than the robin – the wren-tit,[1] a species belonging to a family of birds not known to us in the British Isles. A male wren-tit takes up a territory in the first March after he is hatched, and keeps it all the year round for the rest of his life. He stakes it out and keeps it by song and fight, and when a female joins him she helps in the fighting. If the female dies, the male waits for a new one; if the male dies, the female deserts her territory for, or fuses it with, that of an unmated male. Such territories as fall vacant are taken by males (mainly young) that have none.

It appears that both our definitions of territory (pp. 147–48) work in the case of the wren-tit; judging from their behaviour the pair never falls in sexual activity below a stage equivalent to the period of sexual chases in our standard bird – that is, the male and female *are bound by a sexual bond throughout the year*, though they only reach the stage where the female will accept copulation at breeding time. This behaviour might seem ordinary to a student of monkeys or man, but is quite remarkable among birds.

(2) *Staking the Claim.* – Practically all territorial birds use song, chasing, and fighting in the defence of their plot of ground. In the buntings (which include the American 'sparrows'), behaviour (see Fig. 30A) in defence of the area (see Fig. 33) is almost identical with that of our standard bird. The key to claim-staking is *exhibitionism*; birds make it

1. See Miss Mary Erickson's 'Territory, Annual Cycle, and Numbers in a Population of Wren-tits (*Chamaea fasciata*)' in the *University of California Publications in Zoology*, vol. 42, pp. 247-334 (1938).

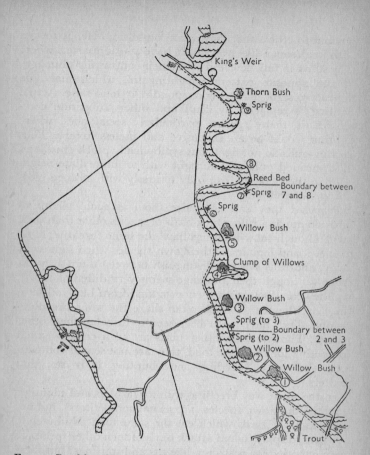

King's Weir

Thorn Bush
Sprig

Reed Bed
Boundary between
Sprig 7 and 8
Sprig

Willow Bush

Clump of Willows

Willow Bush
Sprig (to 3)
Boundary between
Sprig (to 2) 2 and 3
Willow Bush

Willow Bush

Trout

FIG. 33. Reed-bunting territories. The river Isis, between *The Trout* and King's Weir, showing the territories of nine cock reed-buntings, plotted from observation of song-posts. Note how evenly the nine birds occupy the three-quarters of a mile of river-bank, yet how their presence is affected by the bushes and sprigs necessary to sing from. (Based upon Ordnance Survey map, by permission of the Controller of H.M. Stationery Office.)

obvious to others that they are in possession. Hence species
with weak songs, like stonechats, have a headquarters where
they stand in full view and show their contrasting plumage;
at the opposite extreme, nightingales, which make no
appeals to sight, have a correspondingly loud song. Actual
fighting is replaced by bluff and other 'substitute ceremonies' in many species, particularly social ones, where
fighting would be an expensive and useless luxury. Thus
herring-gulls do not fight, but symbolically pluck grass, and
black-headed gulls seldom fight with an immediate neighbour in their colony, but only seriously with strange birds
not known to them.

Provided they can recognize them as such, territorial
birds do not as a rule attack birds of species other than their
own. Though snow-buntings have the same threat reaction
to males and females of their own species when some way
away, and thus *may* not distinguish between them at a distance (although their plumage seems very different to us),
yet they can distinguish their own kind from Lapland buntings or wheatears, which often share the same breeding-
ground. They have never been seen to threaten these other
species. Though these other birds are to a certain extent
competing for the same food, they are not sexual competitors, and to a breeding snow-bunting it is sex that
matters.

In the same way breeding willow-warblers and nightjars
threaten their own species (i.e. sexual competitors), but not
other kinds of birds which eat the same food. Nuthatches,
however, display to and attack birds of such other species as
are competing for nest sites, like tits and starlings; but these
battles are against birds which might interfere with their
breeding-cycle, not their food-supply.

Swans will chase any *large* animal from their territory; although ringed plovers will chase skylarks and linnets, little
ringed plovers (a smaller species) will even try to chase
sheep. We can perhaps explain this by pointing out that
these birds are guarding the nest; ringed plovers, with their

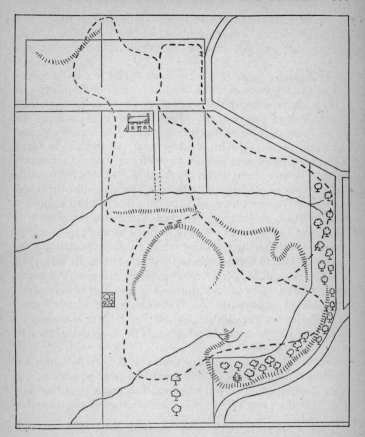

FIG. 34. Territories of horned larks near New York, after Pickwell. The nest is well in the middle of each; note the fact that the territories, though irregular in shape, are roughly equal in area.

nest on the ground, have it exposed to the danger of being trodden on by mistake as well as deliberately raided – not that this explains their attitude to larks and linnets.

We can thus reach the general conclusion that birds normally attack other species in their territory only when those animals tend to interfere with the breeding-cycle.

(3) *Size and Shape of the Territory.* – The normal average size of the territory of non-social birds like our standard one is very constant and of the order of an acre[1] (see Fig. 34). The lower limit of size is nearly always about half an acre, the upper limit may be as much as $4\frac{1}{2}$, quite often under 3, and regularly about $1\frac{1}{2}$. The resident wren-tit and the resident robin have territories always between half and rather over 3 acres. The size is not definitely fixed in all birds. With birds of prey it does not appear to be at all rigid; with others like cow-birds (American parasitic birds) and Dartford warblers, territorialism rather breaks down when the population is high, and in the latter we may find the strange spectacle of several pairs sharing the same territory. Even buntings may have no definite boundaries to their territories. But the constant minimum of half an acre[2] gives us our clue. As previously hinted (p. 151) territories are not indefinitely compressible or expansible. As Professor Julian Huxley says, 'Territories are like elastic disks, of which there is a lower as well as a higher number which can be placed together to cover a given area.'

We can thus reach an important conclusion. Among such birds as hold territories, their practice sets an upper limit to their population. There are two special cases where the half-acre rule does not apply; it is more correct to say this than

1. It is a remarkable fact that the area of Great Britain is 57 million acres, while the May breeding population of land birds is approximately 60 million pairs.

2. The 'half-acre' rule is a proper rule, and not a law. Rules, unlike laws, can be broken. In Africa the minimum territory size for normal land-birds is more like a quarter of an acre, since the thick growth gives birds room in a third dimension. One polygamous bishop-bird holds a territory which is remarkably constant at this size, but another closely related species has territories which are almost indefinitely compressible, and may be under 10 square yards in area.

to say that it is broken. These cases are social birds and 'lek' birds.

Eliot Howard extended his territory theory to cover social birds, explaining that competition for nest sites, which is particularly evident among sea-birds, led such birds to behave territorially even if only in the immediate neighbourhood of the nest.[1] Critics of this attitude, no doubt preoccupied with the food theory, denied that this was true territory. If we accept the definitions on pp. 147–48, we can see that it is. Penguins and gannets tolerate intruders less as these come nearer to their nests. A month or more before the first egg is laid a pair of gannets has taken up its nest site on the cliff, and one of the pair is *always* on the ledge defending it against rival gannets. Any bird within range of the terrible beak, whether it is a gannet or a guillemot, is driven away. Herring- and black-headed gulls and guillemots, while having no definable boundaries round their nests, have a limit of tolerance. Sometimes when these birds are crowded, as guillemots are on top of the Pinnacles in the Farne Islands, every territory is at its minimum size, and the continual bickering, excitement, action, and reaction contrasts strikingly with the comparative peace and quiet of the ledges nearby which are far less thickly populated.

In 'lek' birds the territory is a small plot of ground, held by the male against other males, at a place which has no relation to the site of the nest or the future of the female. Territory is seen here in its most curious, stylized, and symbolic form. 'Leks' are most typical in the cases of blackcocks and ruffs, already described; but meeting-grounds at which males hold territories and display, eventually securing mates, or mating, but which have no other relation (in space) with

1. It is interesting to note that the barn-swallow (the American race of our own swallow), a non-social bird, holds a territory of only a few feet around the nest. Many kinds of swallows are social, and it can be imagined that the barn-swallow is descended from such birds, has become non-social, but has not lost the territory habits it had under socialism.

the breeding cycle, are found also with birds of paradise,[1] with the racket-tailed and other humming-birds, with the cock-of-the-rock, and with many other birds. Such congregations in many cases deserve the name of social displays. Those of the birds we have mentioned are based upon the holding by the males of tiny territories at the meeting-grounds from which they direct their intimidation and posture; and are true territories at that.

(4) *Headquarters, Nest, and Feeding-ground.* – Our standard bird had a few very definite 'headquarters' in his territory, from which he sang and displayed. As time went on he reduced the number of these, showing a definite preference for one or two. This is not always so. Some of the American warblers never have a definite headquarters, but sing from anywhere within the territory. The rosy finch's headquarters seems to be not a bush, a branch, or a bank, but his own female. The broad-tailed humming-bird may have as many as eleven flight-posts in his small territory.

The nest lay within the territory of the standard bird. Even with buntings, which approach the standard very closely, it may not do so. One of the female snow-buntings watched by Dr Tinbergen built her nest outside the male's territory. By dint of a two-day battle with the neighbour, he conquered the ground round the nest. Other buntings (like Bachman's sparrow in America), and even warblers like chiffchaffs, may sometimes build nests outside the territory and often from 100 to 150 yards from the territorial head-quarters. African bishop-birds, which are polygamous, have three or four wives whose nests may occasionally be outside the territory; possibly these birds have a sort of territory (see Fig. 35) which approaches that of the 'lek' birds, which are also polygamous, since, though it is not taken up at a special meeting-ground, it is stylized, symbolic, and a property of the male bird rather than a family affair.

1. It is worth while reading Alfred Russel Wallace's account of the 'sacaleli' or dancing parties of the Greater Bird of Paradise in *The Malay Archipelago* (London, 1886).

Much discussion and controversy has taken place as to the biological value of territory. In particular, the relation between territory and food-supply has occupied many pages of print in the learned journals. Howard was the first to suggest[1] seriously that the system of territory-holding might

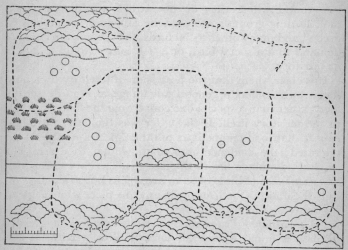

FIG. 35. Territories of four male bishop-birds in Tanganyika. The round circles represent the nests of the females, which may sometimes lie outside the territory, though in this case all lie inside. Male bishop birds are polygamous, and territory seems to be mainly a matter of the male sex-cycle. The small divisions on the scale represent yards. (After Lack.)

confer advantages on birds by assuring them a regular food-supply. Some critics have complained that because birds did not always seek food in their territories, this was not so. It will be remembered that our standard bird sought food for

1. Though Altum in 1868 was the first to put forward this idea on theoretical grounds.

its young outside its territory; as far as I know there is no single case yet published of any species of bird which is *never* known to seek food outside its territory, during such time as it holds one. Warblers, buntings, and dabchicks seek food mostly in their territories. American robins and purple sandpipers feed in them, but also at a common feeding-ground where they do not attack neighbours. Common feeding-grounds are used by lapwings, great crested grebes, and many other social birds. Few people are going to demand that a social bird, or a 'lek' bird, should forage in its territory.

It is clear, then, that the possession of territory by a non-social bird confers a general advantage as regards getting food rather than a special one concerning the food available in the territory itself. If the latter was the case, we might expect that food-competitors of other species would not be tolerated; the evidence goes to show that they are. The general advantage lies in the fact that territory ensures an *even distribution* of birds over an area, thus not stretching unduly the limits of food-supply in any one place while neglecting sources of food in another.

(5) *Courtship*. – True courtship, as distinct from aggressive display, is found among all sorts of birds, territorial and otherwise. The reader is particularly referred to the pictures of courtship and aggressive display in Fig. 30 (pp. 130–33). Its object seems to be to strengthen the bonds between male and female, ensuring that they go through the successive phases of their cycle in a way and at a time suitable to one another, and, more especially, acting as a preliminary to copulation. Courtship may take the forms of vocal demonstration (grebes and divers), sexual flights and chases (buntings, see Fig. 30A), and various ceremonies derived and extended from the ordinary details of daily life, such as the feeding of the female by the male (robins, gulls, and parrots), 'false drinking' (gannets and smews), false preening, wing-lifting (purple sandpipers), short leap-frog flights (green sandpipers), or the presentation or exchange of nest-material.

After the period of sexual flights and preliminary courtship is over, the stage of successful copulation is reached. The act may take place on land, in water, or air, and may follow no apparent ceremony (divers and some terns), or a special courtship display (most birds). In some birds the male may invite the female (gulls, woodpeckers, penguins), in others the female may invite the male (many ducks), either may invite the other (common terns), or it may be difficult to decide which takes the initiative (buntings). Copulation may continue until the first egg is laid (some buntings), until incubation begins (other buntings, herring-gulls, skimmers), or during incubation and even after the eggs hatch (black-headed gulls, common terns).

Detailed descriptions of the courtships of all sorts of birds are now common in bird literature, so much so that the editors of the standard work on British birds (Witherby's *Handbook*) have managed to include a description, in many cases very detailed, of the courtship of *every* important bird on the British list. The present book does not attempt to be a handbook so much as a guide, so I see no reason why I should plague the reader with what has been more than adequately described elsewhere. But though we must leave the appreciation of the individual details of the male-female courtship (as opposed to male-male intimidation) of our British birds to students of the *Handbook* – and the more students it has the better – we cannot leave the subject altogether without giving some examples of how courtship has inspired moving description in bird literature. I will give three such examples:

David Lack, scientist, on the Robin:

The first copulation and the first 'courtship-feeding' occur within a day of each other, but the two are quite separate from each other. The female invites the male to feed her with the same attitude with vigorously quivered wings and loud call as that in which the fledgling begs for food; the male comes with food in his beak and passes it to her, the female swallows it. When the performance starts in late March, the male may just hop up to the female and put food in her mouth whilst she barely quivers the wings or calls, but, after a day or two, both birds are much more excited, the female begging and being fed persistently.

Julian Huxley, scientist, with artistic approach, on the Redshank:

I spent some time watching them, and soon saw the redshanks courting. It was one of the most entrancing of spectacles. Redshanks, cock as well as hen, are sober-coloured enough as you see their trim brown bodies slipping through the herbage. But during the courtship all is changed. The cock-bird advances towards the hen with his graceful pointed wings raised above his back, showing their pure-white under-surface. He lifts his scarlet legs alternately in a deliberate way – a sort of graceful goose-step – and utters all the while a clear, far-carrying trill, full of wildness, charged with desire, piercing, and exciting. Sometimes, as he nears the hen, he begins to fan his wings a little, just lifting himself off the ground, so that he is walking on air. The hen will often suffer his approach till he is quite close, then shy away like a startled horse, and begin running, upon which he folds his wings and runs after. She generally runs in circles, as if the pursuit were not wholly disagreeable to her, and so they turn and loop over the gleaming mud. Then she pauses again, and the tremulous approach is again enacted.

Edmund Selous, artist, with scientific approach, on the Stock-dove:

However it may be, the bow itself – which I will now notice more fully – is certainly of a nuptial character, and is seen in its greatest perfection only when the male stock-dove courts the female. This he does by either flying or walking up to her and bowing solemnly till his breast touches the ground, his tail going up at the same time to an even more than corresponding height, though with an action less solemn. The tail in its ascent is beautifully fanned, but it is not spread out like a fan, but arched, which adds to the beauty of its appearance. As it is brought down it closes again, but, should the bow be followed up, it is instantly again fanned out and sweeps the ground, as its owner, now risen from his prostrate attitude, with head erect and throat swelled, makes a little rush towards the object of his desires. The preliminary bow, however, is more usually followed by another, or by two or three others, each one being a distinct and separate affair, the bird remaining with his head sunk and tail raised and fanned for some seconds before rising to repeat. Thus it is not like two or three little bobs – which is the manner of wooing pursued by the turtle-dove – but there is one set bow, to which but one elevation and depression of the tail belongs, and the offerer of it must not only regain his normal upright attitude, but remain in it for a perceptible period before making another. This bow, therefore, is of the most impressive and even solemn nature, and expresses, as much as anything in dumb show can express, 'Madam, I am your most devoted'.

CHAPTER XI

What You Can Do

✿

So far in this book I have been trying to give the reader samples of things that are being done in ornithology and of the kinds of lives that birds lead. I do not claim that these are the best samples that it is possible to choose; perhaps my own interests have unduly influenced the aspects I have chosen. Perhaps I should regard myself as a sort of commercial traveller, not in boots, braces, or buckets, but in birds. Well, now I have shown you my wares, perhaps you would like to buy something. That is, perhaps you would like to help us get on with the work of finding out as quickly as possible as much as possible about the birds of Britain.

I have hinted in Chaper III at the price you might have to pay for such an interest. It is not very high, and maybe you will find the reward higher. But if you do not want to buy bird watching after what I have said, you will either read one of the better bird books or drop the subject altogether. In any case it is not worth your while to read on. But if you want to become a bird-watcher, here are a few suggestions about what you might do:

(1) Find out if you have a local or district natural history society. If you cannot get hold of the excellent *Directory of Natural History Societies*, published and periodically supplemented by the Amateur Entomologists' Society, it might be worth while to enquire at the nearest public or town library. Possibly the local natural history society might even have notices of meetings displayed in the library. Anyhow, this is the most likely public place. When you have found out about the society, get in touch with the secretary and ask him if you can come along to the next meeting. Do not join till you have

been to a meeting. You can get a line on the sort of leadership and keenness there is. If there is an ornithological section you are lucky. This means that there is an active bird element which is probably working at some practical problem. If such a situation exists, it is worth paying quite a large subscription for the privilege of membership. Fight shy of a natural history society which is a talking shop and which meets practically only to hear lectures. There are many good and honest ornithologists who earn a living by lecturing and showing slides of their photographs, and you will find what they have to say very stimulating. But there are quite a few less well-informed people attendance at whose lectures is just a waste of time.

I am dwelling on this subject because I want people to get the natural history societies looked at from the right point of view. The object of such societies should be, not only the dissemination of knowledge about animals and plants, but also the active increasing of such knowledge. Things are in their right proportion if a keen and positive policy of field work is kept up and the lectures are regarded as the injections of heat necessary to keep the pot boiling.

(2) While you are finding out about the local natural history society, you should make a preliminary study of the birds of your area. Get hold of a county fauna if you can (see p. 51). If there is no book on the birds or general fauna of the district, there may be something in the official county history. The librarian at the Town Hall will know about that. Get about the place on a bicycle. Visit likely places such as woodland, country lanes, thickets, parks, big gardens, rivers, marshes, reservoirs, and sewage farms. But do not let your interest in rarities prevent you from getting a good picture of the general bird life of the district.

It is a good plan to buy a 1-inch Ordnance map of the whole district, a 6-inch map of your local surroundings, and, if you feel like it, a 25-inch map of the immediate neighbourhood. Use these and do not be afraid to mark them. The *Handbook* will very often give you a line on special local problems. Thus, if you live in Berkshire, it is worth getting information about the distribution of the cirl-bunting and the woodlark; information is wanted about the nuthatch in Cheshire, the pied flycatcher in North Wales and the Lowlands of Scotland, the snow-bunting and greater spotted woodpecker in the Highlands, the yellow wagtail in Ireland, and the fulmar on any cliffbound coast.

(3) Decide what special problems you would like to tackle. This may already have been decided for you by the local natural history society, which may be running some survey of its own or helping over one of the national surveys organized by the British Trust for Ornithology or some other central bird organization. Here are some problems that might be interesting:

(a) Establishment of a trapping or ringing station (see Chapter IV).

(b) Preparation of a bird sanctuary.

The organized protection of birds on a large scale has been occupying more and more attention in late years. There are many societies actively concerned with this subject and interested in helping on the rather difficult task of enforcing the Wild Birds' Protection Acts. And without a continual and widespread protectionist attitude towards birds, the watcher of to-day would have fewer interesting things to watch. This book is designed for the individual bird-watcher, as a help to him in that task, and I am trying to avoid a dissertation on the subject of bird politics. And as I am addressing this book, not to large owners of land, but to the many sorts of people liable to buy something for 2s. 6d., it is not suitable that I should suggest any wide plans for the attraction and con-servation of bird life.

But there are plenty of unambitious things you can do, if you have a garden or farm, or if your local natural history society has influence with the Borough Council and an in-terest in the municipal park. The careful disposition of nest boxes and the organization of food-supplies and baths for birds at all seasons may quite easily double the bird population of a residential area. At Whipsnade Zoo, where this kind of thing is done, the population in the Woodland Bird Sanctuary is about five times the normal population for wild woodland of the same type.

The measures you can take are in the main four. First you should provide additional nest sites. You would do best to make your nest boxes yourself; they can seldom be bought cheaply enough, because of high distribution costs. There is no need to go in for fancy pieces of bark or other fussy decoration. Boxes should be simple constructions of creosoted deal and should have a hinged roof sloping downwards and forwards, over which it is a good idea to tack a piece of roofing felt. They should be of minimum measurements 6-inch cube inside and should have a hole of a size suitable for the intended occupant. The best holes are the size of a two-shilling piece for tits, half-crown size for larger tits and nut-hatches, and larger for starlings, jackdaws, etc. I do not much fancy the type of nest box hollowed out from the bole of a birch tree, though they are often very effective. If the bottom of the hollow part is left hemispherical, the young birds tend to slide down into the middle and among a brood of perhaps ten tits the bottom ones are pretty certain to be crushed.

Secondly, there should be as much cover as possible for your birds. By cover I mean variety of plant growth. A wilderness is to be encouraged from the ornithological point of view, provided the trees, shrubs, and bushes in it are of different shapes and sizes, presenting as much total surface area as possible.

Thirdly, water is an essential. Birds need it for drinking and bathing, and they need it all the year round. Some people fix a permanent water supply for their birds by having an electric bulb or some other means of heating sunk in the concrete at the bottom of their bird bath. This means that, when everything is frozen solid in winter, the birds can always get a drink. And this is not as expensive as it sounds; it might cost 10s.

Fourthly, you should feed the birds. I am not suggesting

that you should feed them all the year round or that you should spend a great deal of money on doing this. It is a pure waste to put out grain or bread in the summer when the birds are busy seeking insects for themselves and their young. On the other hand, if you get a winter which is as severe as that of 1939–40 or that of 1946–47, you are doing a great service to humanity as well as to the birds if you give them all you can spare.

In winter any spare grain, bread, and fatty things like nuts and suet are greedily eaten by the birds, and by giving them concentrated nourishment in a very short time you enable them to stoke up enough in the short hours of daylight to keep themselves going through the long cold nights. The B.B.C., I am glad to see, announces in its news bulletins in hard winters that people would be well advised to feed the birds, and though this draws down criticism from various people, who are ill-informed about the habits of small birds, it certainly has an excellent effect. My friend Tom Harrison once asked in a broadcast why people fed birds. He got letters back from well over a thousand people who did; the reasons they gave ranged from 'my father always did' to 'it gets rid of the scraps'. But the net result was most encouraging. If there are over a thousand people who (i) feed birds, (ii) listen to bird talks, and (iii) having listened write letters about the question, there must be an incredible number of bird feeders in our country.

(c) Armed with a trapping station and a bird sanctuary of a sort, you are in a position, with your garden or whatever it is, to study further problems. It should not take you long to know every individual bird in the place, particularly as you will be certain, after a very few months, to have ringed every bird in the garden, and you can help your knowledge of individuals by putting coloured rings on their legs also. I suggest, then, that you make a study of a particular pair of birds, from the taking up of their territory to the dispersal of their young. A great deal is known about robins, and a fair amount about blackbirds. Perhaps chaffinches would offer

you some scope for new discoveries, and certainly an un-
usually small amount is known about the domestic habits of
the house-sparrow. The British Trust for Ornithology issues
(and collects) excellent cards on which the history of observed nests can be recorded for scientific analysis.

(*d*) You might start some behaviour experiments, using
stuffed mounts set up within territories, near nests and so on.

(*e*) Make a chart of the days on which the various species
of birds you can hear are singing, and keep this song chart
up-to-date all the year round. Recently the British Trust for
Ornithology has organized a nation-wide survey of the song
periods of various common British birds.

(*f*) Choose one individual species of bird and make a
special study of its distribution in your locality, find all the
nesting places, and make detailed notes on habitat and site,
as well as on simple geographical position. If this can be
combined with an exact census of nests over a largish area,
so much the better.

(*g*) Try to find every nest within a quarter of a mile of one
particular spot. Correlate these nests with the singing birds
and examine the pattern you get in the light of the territory
theory (see Chapters IX and X).

(*h*) Co-operate personally, or get your natural history
society to co-operate with the British Trust for Ornithology
in any of its surveys. Recent surveys have been nation-wide
investigations into the distribution of the corncrake, the fulmar, the woodcock, the bridled guillemot, and the black redstart, investigation of the habits, brood size, and breeding
biology of the swallow, analysis of the food of the little owl,
the origin of the habits of milk-bottle opening and paper-
tearing by tits, a study of tameness and an investigation of
bud-eating, enquiries especially designed for local natural
history societies into the regional distribution of birds like
the curlew, teal, and sand-martin.

(*i*) Prepare a tally-list for your area (see p. 58).

(*j*) Make comparative walks through woodland, moorland, and agricultural land, and note the number of times

different species are met with. In your notes compare the
number and species of birds in these three different habitats.

(*k*) Make a trip to a sea-cliff where birds are breeding or
to a mud-flat where waders and ducks assemble. This need
have no purpose other than that of acquiring atmosphere,
and indeed any special bird trips to good places are a wonderful stimulus for the carrying on of the routine work which
is really important.

(*l*) On your way to your place of holiday, or on any journey, do transects. These can be effectively made from train,
car, bicycle, or ship. Note the different species you see, the
kind of country you see them in, and their actual numbers.
If you cannot make accurate counts, try to keep an index of
relative numbers. Sea transects are very useful things to keep.
If you are on a long sea journey, get hold of the log so that
you know the noon position of the ship each day, and mark
the birds seen along the ship's course. It is a good idea to
come on deck at two-hour intervals during daylight and
note all the species and numbers of birds visible at one time.
For the identification of sea birds you cannot do better than
get W. B. Alexander's *Birds of the Ocean* (New York and
London, 1928, and later editions).

(*m*) If your interests extend to what is going on in the
whole country rather than in the local field, join a national
organization. There are quite a number. The senior body,
or what might be better described as the learned body, is the
British Ornithologists' Union. New members of this have to
be proposed and vouched for by an existing member and
seconded by another member or members. The B.O.U.
publishes the senior bird journal in Britain, the *Ibis*. This
appears quarterly, usually runs to over 1,000 pages a year,
and deals with birds from every part of the world. There are
many papers about Africa in it and a large number deal
with classification, evolution, behaviour, and other subjects.
Though it is a very necessary and important journal for the
advanced scientific ornithologist, there is little in it of
interest for the absolute beginner in British birds.

A sort of inner circle of the B.O.U., largely composed of London members, goes by the name of the British Ornithologist's Club. Every month, except in the late summer, this club meets, and communications, verbal or written, are submitted to the meetings. They may be about any subject of current interest and are published in a *Bulletin* issued shortly after each meeting.

The chief organization concerned with field work in Britain is the British Trust for Ornithology. This was founded in 1933 and incorporated in 1939. Its main object is to secure the efficient collection of information about British birds, and to this end it promotes some of the special enquiries I have already mentioned. Some of the work has been done by supporting the University of Oxford, under whose direction is the newly-established Edward Grey Institute of Field Ornithology. The Trust is rather a new development in ornithology, for it is a private concern run by voluntary labour. It is doing the sort of work which in some countries is done by Government Departments like the Wild Life Service in America. Most of its energy comes from its individual members and from such local natural history societies as have agreed to affiliate and help over enquiries. There is a form for membership of the Trust at the back of this book.

If you are interested in bird protection, the senior body is the Royal Society for the Protection of Birds. This very old-established society (it was founded in 1889) has a great deal of support and does much, particularly in education and propaganda, in the managing and financing of many bird sanctuaries and their watchers, and in the payment of bounties on the successful breeding of rare birds. This was a task which was previously undertaken by the Association of Bird Watchers and Wardens; but this association, which did much good work, has now 'merged' with the R.S.P.B. The R.S.P.B. works closely with such regional bodies as the West Wales Field Society over particular problems – such as the protection of our last remaining kites. There is a form for membership of the R.S.P.B. at the back of this book.

If you are interested in bird photography, the suitable national organization to join is the Zoological Photographic Club. One of the tasks of this club has been the accumulation of a national standard collection of bird photographs, with the idea that a copy of every good photograph of every species should be kept in some permanent place of reference.

Some people like to make paper plans and push them through regardless of distraction. Others like to wander into a subject and let it take them where it will. I hope I have been able to give the paper-plan people some preliminary headings and I hope that for the others I have given the stream of ornithological thought enough momentum for it to dislodge them from their rest. Bird-watchers are in some ways like golfers, who want to convert all their friends to golf. I do not excuse myself for belonging to this class. If this book succeeds in making new bird-watchers, I hope they will have a very happy time, though I do not see how they can help it.

Index

To the Secretary
The Royal Society for the
  Protection of Birds
82 Victoria Street
London, SW1

Having read James Fisher's *Watching Birds*, I would like to join the Royal Society for the Protection of Birds, and enclose a subscription (the Society asks its Fellows to pay £1 1s. a year – or more if they can – its Members to pay at least 10s. a year – or at least 5s. if they are under 21).

I am interested in the Society and would like some literature about it.

Name ....
(MR, MRS, MISS, ETC.)

Address..

....

.... *Date*........

PLEASE WRITE IN BLOCK CAPITALS AND CROSS
OUT WHAT DOES NOT APPLY

ed. 1951 W.B.

THE ROYAL SOCIETY
FOR THE PROTECTION OF BIRDS

This old-established society (it was founded in 1889 and incorporated under Royal Charter in 1904) was originally started to combat the plumage trade, which involved barbarous cruelties to egrets and other beautiful birds. Now its main efforts are exerted to increase the number and variety of British wild birds; to preserve all that are rare, beautiful or scientifically interesting; to prevent any sort of cruelty to wild birds – deliberate or otherwise; and to influence and educate public opinion about birds.

To these ends it:

Owns or manages many sanctuaries in the British Isles, where birds which are in special need of protection often nest in surroundings of great beauty;

Supports generally the conservation of wild animals and plants in their natural communities, and particularly the preservation of natural marshland, mountain and moorland, and coastland;

Upholds, and strives to improve, the Wild Birds Protection Acts, under which the taking or killing of most British birds and their eggs is unlawful;

Spreads the knowledge and love of birds among old and young by means of films, lectures, meetings, exhibitions, posters and competitions (in particular by open essay competitions and by a Bird and Tree Festival Scheme for schools);

Organizes, in collaboration with the British Trust for Ornithology (see overleaf) the Junior Bird Recorders' Club for boys and girls between the ages of fourteen and eighteen;

Publishes the quarterly magazine *Bird Notes* (free to Fellows and Members), and many reports and pamphlets.

There is a tear-off form for Fellowship (or Membership) on the previous page.

Note: The R.S.P.B. (overleaf) and the B.T.O. (opposite) each have quite different kinds of work to do. They are in no possible sense rivals or competitors, but often collaborate *e.g.* in the exchange of information.

IF YOU WANT TO USE THIS SHEET, CUT ALONG
THE DOTTED LINE

To the Secretary
The British Trust for Ornithology
91 Banbury Road
Oxford

Having read James Fisher's *Watching Birds*, I would
like to join The British Trust for Ornithology, and
enclose a subscription (anybody over 17 may apply
to join: the Trust asks ordinary members to pay £1 a
year, or more if they can, but those between 17 and
21 may pay 10s.).
I am interested in the Trust and would like some
literature about it.

Name ....
(MR, MRS, MISS, ETC.)

Address ....

....

.... *Date*..........

PLEASE WRITE IN BLOCK CAPITALS AND CROSS
OUT WHAT DOES NOT APPLY

ed. 1951 W.B.

THE BRITISH TRUST FOR ORNITHOLOGY

This relatively new organization (it was founded in 1933 and incorporated as a Company in 1939) was started for the purpose of increasing our knowledge of British birds, by means of observation and experiment in the field, *i.e.* out of doors.

To this end it:

Has successfully promoted the establishment of the Edward Grey Institute of Field Ornithology at Oxford, which is already the recognized centre of scientific study of the living bird in Britain;

Organizes and supports enquiries and investigations, some co-operative, some individual: among those already undertaken and reported upon are enquiries on the life-history of swallows, the distribution of woodcock, the food of the little owl, the numbers of the black-headed gull, the spread of the fulmar, the numbers and economic value of rooks and woodpigeons, the distribution of the 'bridled' form of the common guillemot, the history of the corncrake, the opening of milk-bottles by tits, the spread of the black redstart, the migratory movements of swifts, waders and terns, and the nesting population of great-crested grebes;

Promotes annual investigations of the nesting of all common birds (the Nest Records Enquiry) and of the numbers of herons;

Administers the national scheme for marking wild birds with numbered rings to provide information about their migrations, longevity and other problems;

Co-ordinates the scientific activities of the several Bird Observatories on the coasts of Britain, established for the study of migration;

Holds many scientific meetings, exhibitions and discussions in different parts of Britain;

Maintains the Alexander Library at Oxford, jointly with the University (a useful range of publications can be borrowed by members through the post);

Gives information and advice to bird-watchers on useful subjects for research and investigation both from its office and by means of Regional Representatives all over the British Isles; also considers applications for small grants in aid of research by groups or individuals;

Publishes frequent *Bulletins*, annual *Reports*, occasional field guides and circulates a summary of *Recent Publications in Bird Biology*.

There is a tear-off form for Membership on the previous page.